Seeds of Insurrection

Seeds of Insurrection

Domination and Resistance on
Western Cuban Plantations, 1808–1848

MANUEL BARCIA

LOUISIANA STATE UNIVERSITY PRESS ✦ BATON ROUGE

Published by Louisiana State University Press
Copyright © 2008 by Louisiana State University Press
All rights reserved
Manufactured in the United States of America
FIRST PRINTING

Designer: Barbara Neely Bourgoyne
Typeface: Seria
Printer and binder: Thomson-Shore, Inc.

Library of Congress Cataloging-in-Publication Data
Barcia Paz, Manuel, 1972–
  Seeds of insurrection : domination and slave resistance on western
Cuban plantations, 1808–1848 / Manuel Barcia.
     p. cm.
  Includes bibliographical references and index.
  ISBN 978-0-8071-3365-1 (cloth : alk. paper) 1. Slavery—Cuba—
History—19th century. 2. Slave insurrections—Cuba—History—19th
century. 3. Plantations—Cuba—History—19th century. I. Title.
  HT1076.B376 2008
  306.3'62097291109034—dc22

                                        2008026516

For Effie

Contents

Illustrations

Acknowledgments

It has been a long road since I took the belated decision to attend university in Havana some fifteen years ago. Many people have come to and gone from my life during this time. Before I begin the long list of acknowledgments that will follow, I would like to say that I am deeply grateful to all those who helped me in one way or another and who will not be mentioned here due to reasons of space or bad memory.

I would like to start by recognizing the continuous encouragement I received from professors and fellow students at the University of Havana. Many of them have had a lasting impact on my professional and personal maturity. At the Museum of the City of Havana, I had the opportunity to spend nine amazing years working with the most professional and kind-hearted people I have ever met. Special thanks go to Roger Arrazcaeta, Carlos Alberto Hernández, Dania Hernández, Karen Mahé Lugo, Sonia Menéndez, Anicia Rodríguez, Arelys Hernández, Lissette Roura, and Francis Sera, all in the Gabinete de Arqueología. Thanks go as well as to the personnel of the Museum's archive and library, and especially to the Historian of the City, Eusebio Leal Spengler, for giving me access to his encyclopedic historical knowledge every time I needed it and, more importantly, for considering me not only a colleague but also a friend.

The staff of the Archivo Nacional de Cuba greatly facilitated the final outcome of this research. They were constantly at hand to answer my endless queries and to resolve my frequent complaints. My special gratitude to Julio López, Barbara Danzie, Jorge Macle, Isabel Meriño, Marlén Ortega, and Isabel Oviedo. The same applies to the staff of the provincial archives of Santiago de Cuba, Matanzas, and Pinar del Río, and to the personnel at the National Archives in London.

The Department of History and the Centre for Latin American Studies at the University of Essex became a sort of second home for me over the past years. There I had the chance to learn from colleagues and friends on a daily basis. My particular appreciation goes to Chris Anderton, Monika Baar, Robin Blackburn, Andrew Canessa, Kim Coleman, Jane Corbey, Catherine Crawford, Jim English, Vic Gatrell, Paul Glenister, Jeremy Krikler, Malcolm McLaughlin, Mark Merry, Myra Offord, Christina Schröder, Erna Von der Walde, Belinda Waterman, and Lisa West.

I wish to extend a special note of gratitude to my colleagues and friends at and around the University of Leeds, where I have been welcomed with open arms and given the opportunity to develop my ideas within the newly established Institute for Colonial and Postcolonial Studies and the Department of Spanish, Portuguese, and Latin American Studies. My sincere thanks to Nir Arielli, Ben Bollig, David Braithwaite, Francesco Capello, Vania Celebicic, Karen Charlesworth, David Frier, Stuart Green, Bettina Hermoso, Graham Huggan, Antonio and Hayley Martínez Arboleda, Brendon Nicholls, Pablo San Martín, Andrew Thompson, Jutta Vetter, Luna Zabalza, and especially to Paul Garner and Louise Gibbs, who hosted and took care of me and my wife just after our arrival in Leeds, and this despite my repeated jokes about their home-grown courgettes.

Across the world I was always able to resort to other friends and colleagues who gave me the moral support and professional encouragement I frequently needed. Exceptional credit to Ileana Alvarez, Ana Andreia Araújo, Janina Arsenjeva, María Eugenia Brito, Mariveliz Cabán, Carmen Cabrera, Michelle Chase, Jonathan and Eldimarys Curry-Machado, Katia Enrique, Dan Entwistle, Reynaldo Funes, Tom Goddard, Dominique Goncalves, Nurcan Kaya, Eleni Kokkinou, Leo Kolenkine, José Guadalupe Ortega, Ricardo Quiza, Yazmín Reyes, José R. Rivera, Aymée Robaynas, Arlé Valdés, and Luzmira Zerpa.

I am also indebted to a large number of scholars I met in a way or another, mainly at conferences and archives in Latin America, Africa, or Europe. Long chats and discussions helped to broaden my understanding of the issues I discuss in these pages. Thanks to Ulbe Bosma, Ada Ferrer, Gloria García, Carlos A. Forment, Alejandro de la Fuente, David Geggus, Lillian Guerra, Aline Helg, Walter Johnson, Jane Landers, Javier Laviña, Paul Lovejoy, Rafael Marquese, Luis Martínez-Fernández, Marc McLeod, Andrew Mc-Michael, Joseph C. Miller, Sidney W. Mintz, David Murray, Robert Paquette, J. D. Y. Peel, Reynaldo Román, James C. Scott, Samita Sen, Mimi Sheller, Elisée Soumonni, Dale W. Tomich, Pablo Tornero, Jim Walvin, Amanda Warnock, David Wheat, and Michael Zeuske.

This study draws heavily upon African history, a notoriously difficult field, and I have been privileged to receive insightful comments of two of the most important contemporary scholars in African history. Both John Thornton and Robin Law gave me their time and expertise, answering every one of my many questions within hours. Their contribution to this book is invaluable.

Matt D. Childs has been a great friend throughout the years; he has always been ready to talk about slavery in Cuba. I would later regret not to mention here our long—seemingly endless—conversations about African slavery in the backyard of La Columnata Egypciana café in the heart of the Old Havana. Adrián López's vision of the history of Cuba has influenced my own in an enormous extent. Although we have always had disagreements, we have managed to pull through difficult times together, giving each other support and encouragement—things we both considerably needed during our time in Cuba.

María del Carmen Barcia has been deeply involved in my life since that lucky morning when I approached her at the National Archive in Havana. Ever since then she has kept a motherly eye on me, looking after my professional development and giving me advice about the good and bad of this world. I will be always thankful to her for never losing her faith in me. After all, I am nothing but her Frankenstein—I even carry her surname!

Mary Ellen Curtin also deserves special recognition. I will always remember our endless chats in the Blues Bar over cups of espresso. I greatly benefited from her sound knowledge of African American history, her always precise critiques of my work, and her excellent sense of humor.

Matthias Röhrig Assunção took me under his wing and guided me through the complexities of the British university system, habitually offering encouragement and getting involved in this project almost as much as I did. As his Ph.D. student, I was able to learn the trade and to have fun at the same time. I will always treasure his friendship and will miss his terrific feijoadas.

My editor Alisa A. Plant and the staff at Louisiana State University Press have done a great job from the start. Alisa in particular has been supportive throughout and has helped me clarify many doubts. I have benefited from her expertise, professionalism, and insightful comments.

My Greek in-laws, Pavlos, Kiriaki, and Giorgos, were an incessant source of help and tenderness. They hosted me in their house periodically, graciously coping with my complaints and poor knowledge of the Greek language. I am trying to reduce the former and to improve the latter.

My parents, Elvira and Juan Manuel, my sister Celia, her husband Orestes, and my nephew Bryan are always with me, no matter how far apart we are in geographical terms. I owe them so much that it is practically impossible to articulate my debt in words.

Many things have changed in my life throughout these years. None, however, has had a bigger impact than meeting my wife, Effie Kesidou. Today I am a happy man—perhaps the happiest of men—because of her. She has been permanently on my side, in good and bad times alike. I thank her for being patient and tolerant and for giving me her love unconditionally. Not surprisingly, then, this book is dedicated to her.

Seeds of Insurrection

Introduction

The seeds of the insurrection have been profusely spread.
—Apolinar de la Gala, prosecutor's conclusion, causa 38, 1844

In March 1843, hundreds of African slaves rose against their masters and overseers in the western Cuban district of Bemba. After killing a large number of people and creating panic all over the island, they were finally seized and massacred. In the aftermath of the revolt, some survivors who were still at large committed suicide, while others were apprehended and sent back to their plantations. This violent slave uprising and the plot that was behind it, by means of which the conspirators created an opportunity to fight for their freedom, constitute only one among many acts of resistance recorded by Cuban colonial authorities during the nineteenth century. Cuban slaves were subjected to a system that relied on unambiguous practices of domination, and they did not always accept this domination passively. Far from being quiescent, they tested the limits of the institution of slavery in a wide variety of ways.

Day-to-day life on western Cuban plantations throughout the first half of the nineteenth century was shaped by diverse forms of repression and a constantly evolving process of acceptance and transformation. In every one of the New World societies to which they were transplanted, recently arrived slaves and their descendants created spaces in which to keep alive their own cultures. They made use of the privacy offered by their huts and barracks, of the lack of surveillance in the cane and coffee fields, and the anonymity of the roads, taverns, and shops to voice their ideas and opinions with an acceptable degree of safety. The western lands of Cuba, brimming with coffee and sugar estates, were no exception. Colonial authorities and slave masters often permitted the slaves their spaces of culture and expression as safety valves in the interest of maintaining social stability. Alterations in the precarious balance between slaves and masters could lead to violent acts of slave resistance.

In this book, I consider the different forms of resistance practiced by African-born slaves and their descendants under Spanish rule in Cuba, using as my primary sources the numerous cases of slave resistance recorded by the Spanish colonial authorities. Although this is not the first work to assess the varieties of slave resistance in the Americas, it is the first to do so for the specific case of nineteenth-century Cuba. This book is also the first to consider the origins of the slaves in Cuba as an integral part of the story of their resistance. In doing so, I reveal that slaves were neither simply spectators to the events that surrounded them nor happy participants in their own oppression. Instead, they resisted domination in its countless forms by negotiating, by reproducing their cultures, by openly revolting, by running away to the forests and mountains, and by taking their own lives. More importantly, slaves resisted domination in ways that accorded with their personal life experiences. For those born in Africa, the process of resettlement was a cultural shock that involved learning a new language, new social practices, and a new life as a slave. As Chapter 2 demonstrates, most of the slave revolts of the nineteenth century had African-born leaders who pursued what Eugene Genovese has called "restorationist aims."[1] The Creole descendants of the first generations of African slaves, in contrast, more frequently had recourse to Spanish colonial laws that conceded rights to those slaves who were able to realize that they were entitled to enjoy them.[2]

This is not to say that Creole slaves did not revolt or that African-born slaves did not take advantage of the law. But the attitudes and forms of resistance of Creole and African-born slaves differed. Creoles had lived their entire lives in direct contact with Spanish Cuban culture. Most of them learned both the language of their ancestors and Spanish. They were children of the land in which they lived, and consequently they knew its secrets.

African-born slaves, in contrast, never relinquished their African heritage. They carried with them their memories, their cosmologies and religious beliefs, their jokes and songs, their knowledge of war and politics, their codes of honor, and many other beliefs and values that defined who they were and how they viewed the world. As John K. Thornton has pointed out, "Whatever the brutalities of the Middle Passage or slave life, it was not going to cause the African-born to forget their mother language or change their ideas about beauty in design or music; nor would it cause them to abandon the ideological underpinnings of religion or ethics—not on the arrival in America, not ever in their lives."[3]

Approaches to Slave Resistance in the Americas

New World slave regimes and the forms of resistance developed to fight them have been a matter of academic interest since the late nineteenth century. Yet scholars became more engaged with these subjects only in the first quarter of the twentieth century. This period was dominated by the myth of the happy and docile slave who was easily integrated into his or her new environment. Renowned scholars reproduced ancient racist assertions about the inferiority of blacks and especially about their criminality-prone personalities.[4] Some authors, including Fernando Ortiz, Ulrich Bonnell Phillips, and Gilberto Freyre, characterized slaves as inferior people who believed in and practiced various types of witchcraft. Public opinion gave credibility to these viewpoints, which reflected the influence that the racial conflicts of the early twentieth century in the U.S. South and Cuba exercised on scholars.[5]

It was only after 1940 that the scholarship of slavery and slave resistance began to change. Two studies played an important role in inspiring the reassessment of slave behavior. The first was a 1941 article by Raymond and Alice Bauer, in which they attacked the myth of the passive slave and suggested for the first time the concept that slaves resisted their oppression in nonviolent, everyday forms. The Bauers researched the patterns of day-to-day slave resistance in the antebellum South, noting that slaves deliberately slowed down work, destroyed their masters' property, and even practiced self-mutilation as valid forms of resistance.[6] In 1943, a book with a more political character definitively changed the direction of the historiography of slavery. Herbert Aptheker's *American Negro Slave Revolts* challenged the established historiography by arguing for existence of a revolutionary tradition among African

slaves of the antebellum South. Aptheker drew evidence to support his argument from best-known conspiracies and revolts of the nineteenth century,
which included Gabriel Prosser's plot of 1800, Denmark Vesey's conspiracy
of 1822, and Nat Turner's rebellion of 1831.[7]

Despite its merits, Aptheker's book underestimated the importance of
acts of day-to-day resistance and overestimated the importance of marronage and slave revolts, a bias that has characterized subsequent histories
of slave resistance, including those by José Luciano Franco on Cuba and
C. L. R. James on Haiti.[8] From the late 1960s onward, scholars began to pay
attention only to those forms of resistance that could be included under
the umbrella term of "slave rebellions." Despite the efforts of a few scholars,
including Monica Schuler and Stuart Schwartz, to turn this tide, the debate
surrounding slave revolts in the New World absorbed the attention of most
scholars working in the field.[9] Discussions of slave resistance for the most
part considered as such only marronage in specific regions (such as northeastern Brazil or Jamaica), conspiracies, and rebellions.

The academic debate about the character and aims of slave rebellions in
the New World has been most fiercely fought in the last decades. Eugene
Genovese claimed in 1979 that "by the end of the eighteenth century, the historical content of the slave revolts shifted decisively from attempts to secure
freedom from slavery to attempts to overthrow slavery as a social system."[10]
In other words, he argued that the vast majority of slave rebellions prior to
the 1790s had an intrinsically escapist ("restorationist") character, while most
rebellions that took place afterward were devoted to eradicating slavery as an
institution and were inspired by "bourgeois-democratic" politics. Although
his work focused on the Old South, Genovese erroneously extended this
"shift of character" to other slave societies in the Americas, ignoring the specific histories of each region. This prompted an international discussion
that took place over the course of the 1980s about the character and causes
of New World slave uprisings.[11]

Although Genovese took care to specify that he did not intend to suggest
"the disappearance of restorationist revolts at any point in time," many other
scholars judged his argument to be at the very least reductionist. Two lines of
argument developed in answer to Genovese's hypothesis. The first was generated by scholars whose studies focused on the British West Indies, where
an early process of Creolization—early in comparison to other territories of

the New World, such as Brazil or Cuba—took place as a consequence of the abolition of the transatlantic slave trade in 1807.[12] Scholars focusing on the history of Africa presented a second approach that considered the continuity of African cultures in the Americas to be the most significant influence on slave revolts.[13]

Although historians of the 1980s prioritized the study of violent forms of slave resistance, the abolition process, and the transatlantic slave trade over other topics, nonviolent forms or resistance found their way into some publications. In their analysis of two of the most important slave movements in Latin America during the nineteenth century, Robert Paquette and João José Reis paid attention to the period that foreshadowed every revolt, during which plots were planned and organized. By investigating the daily lives of the common people involved in these movements, Paquette and Reis provided fresh insights into slaves' private and public worlds.[14] In books of more comprehensive scope, Michael Craton and David Barry Gaspar examined daily life in the slave societies of the British West Indies and Antigua. Both Craton and Gaspar discussed the roles played by slaves and the identities of their leaders within these societies. They also explored the significance of slaves' backgrounds and to what extent those backgrounds determined or influenced their resistance.[15]

Studies on the forms of resistance practiced by subordinate groups became more popular after the emergence of the subaltern studies school of thought and the publication of books by James C. Scott in the 1980s and early 1990s.[16] Historians of slavery in the Americas reacted immediately to Scott's works. Emilia Viotti da Costa, for example, made use of Scott's concept of the hidden transcript in her remarkable study of the Demerara slave uprising of 1823. More critically, Robert L. Paquette questioned Scott's perspective on African slaves' day-to-day resistance in the New World. He argued that for Scott and his followers, "leadership counted for little" and "distinctions between political and apolitical resistance blur."[17] Today, historians of African slavery in the Americas hold contradictory opinions about Scott's ideas, but Scott's work has stood the passage of time. By looking at what many had failed to spot before, Scott has given new and fresh inspiration for the further study of forms of domination and slave resistance in the New World, prompting scholars to recognize more openly that those slaves who did not resist their bondage in a violent manner were far from

being passive, agreeable witnesses to their subordinate condition.[18] In order
to understand these other rebels, Scott's approaches—and in particular his
concept of the hidden transcript—are extremely useful.

Cuban Slavery and Historiographical Endeavors

Cuban historiography did not escape the dominant trend of focusing on
specific forms of slave resistance, such as marronage and rebellion, while
leaving others— no less important—in the shadows.[19] In the first half of
the twentieth century, Cuban historians and fiction writers produced only a
few narratives about the history of slavery.[20] In the same period, however, a
debate about Cuban culture was burgeoning. Fernando Ortiz unquestion-
ably made immense contributions to this debate. His numerous studies and
publications blazed a trail for Afro-Cuban studies in the first half of the century. Ortiz assessed the cultural contributions of the different ethnic groups
that arrived in Cuba and tried to identify their places of origin in Africa.
He showed great interest in music and folklore, religious ceremonies, and
practices of mutuality and association. Some of his books remain classics
today. Ortiz also introduced the concept of transculturation, used today with
enthusiasm by scholars of postcolonial studies.[21]

Although Ortiz was almost alone in his pursuit of the African roots of
Cuban culture, other Cuban scholars made significant contributions to this
theme from diverse angles. José Luciano Franco, for example, published an
interesting but not particularly innovative book about Afro-Cuban folklore
in 1959. In the early 1960s, Pedro Deschamps Chapeaux began publishing
articles with a transparently ethnographic character. In the 1960s and 1970s,
Deschamps Chapeaux was virtually the only scholar in Cuba writing about
Afro-Cuban culture. In Miami, however, Lydia Cabrera followed Ortiz in
directing attention to a broad range of topics concerning Afro-Cubans, from
forms of ritual magic to culinary recipes.[22]

In the study of African slavery, Raúl Cepero Bonilla, Manuel Moreno
Fraginals, and other social scientists initiated a new trend in the 1950s that
reached its peak with the publication of the first volume of El ingenio in 1962.[23]
This promising moment in the analysis and understanding of the forms of
colonial domination and slave resistance was cut short for political reasons,
however: the Cuban Revolution of 1959 drew upon the "glorious past" of the
nation to legitimize itself, and the exterminated aboriginal inhabitants of

the island, African slaves, and mambises (the name given to patriots during the wars of independence from Spain) were all included in the revolution's pantheon.[24] From 1959 onward, Cuban scholars of slave resistance focused overwhelmingly on slave rebellions and marronage, while other forms of slave resistance were largely overlooked. The prevailing political discourse offered an analogy between Cuban revolutionaries and the rebel slaves who died fighting against their masters over the course of almost four hundred years of colonial domination. After the revolution, Cuba—for a long time the only socialist country in the Americas—portrayed itself as a sort of maroon island defying the will of the master, the United States of America.

In the wake of the revolution, many scholars of slavery and Afro-Cuban history, including Fernando Ortiz, began writing books on totally different topics.[25] Others, such as José Luciano Franco and Juan Pérez de la Riva, opted to sail with the wind. Franco and Pérez shifted their scholarly attention to the field of the economic history, which was favored by the new regime, though Franco did write some books about slave rebellions in Matanzas and Santiago de Cuba.[26]

Though revolutionary policies constrained academic work on slavery in Cuba in the 1960s and afterward, scholars did not fail to notice that Cuban slave owners had often adopted violent practices as a way of ruling and controlling their slaves. As Reis and Silva have rightly noted regarding nineteenth-century Salvador da Bahia—a quite similar slave society to Cuba's—"the owners and slave authorities in Bahia, as everywhere, made use of violence as a fundamental method for controlling the slaves." They also pointed out that "the fight against slave autonomy and lack of discipline, while working or not, was made through a combination of violence with negotiation, of whip and reward."[27] In recent years, a few studies have appeared on slavery and the law in Cuba and other colonies of the New World. These works focus on how the colonial apparatus of domination intended to establish its rules through negotiation.[28] Studies of the relationship between the Cuban colonial government and the slaves have appeared since the 1980s. These works remark on the importance of the social changes and new ideologies that appeared in Cuba as a consequence of the American, French, and Haitian revolutions.[29]

In Cuba today, studies of marronage and slave rebellions continue to predominate over scholarly explorations of other types of slave resistance.[30]

The exception is Gloria García's *La esclavitud desde la esclavitud: La vision de los siervos* (*Slavery from Slavery: The Servants' Vision*).[31] In this book, García offers a model for future studies of domination and slave resistance in Cuba. She analyzes forms of slave domination, commenting on how these forms changed over time as a result of both internal and external factors, and she evaluates forms of slave resistance. Although García pays attention to plots, revolts, and marronage, her main contribution to scholarship is her assessment of the many other forms of slave resistance that Cuban scholars had virtually ignored. Discussions of suicides, homicides, and the diverse ways in which slaves made appropriate use of hegemonic channels in order to achieve their freedom or other goals appeared for the first time in García's book. Only recently published in Cuba, it constitutes a valuable attempt to establish new avenues in the study of domination and slave resistance on the island. Some other studies on broader or narrower subjects have since been published, but García's work continues to play a significant role in the development of new ways of looking at slave resistance in Cuba.[32]

Since the first half of the twentieth century, scholars from the United States have also contributed to the study of slavery in Cuba, and such work has become increasingly prominent since the triumph of the revolution led by Fidel Castro in 1959. Gwendolyn Midlo Hall and Herbert Klein compared slavery in Cuban society to slavery in Saint Domingue and Virginia.[33] Franklin Knight, Verena Martínez Alier, David Murray, Rebecca Scott, and Robert Paquette assessed Cuban slaves' kinship structures, forms of resistance, and their most important conspiracies and rebellions.[34] Such works have shed light upon matters crucial to understanding Cuban slavery, and simultaneously have introduced new approaches to rival the Cuban government's official history. Martínez Alier's book, for example, was a pioneering work in the field of gender studies, while Scott was the first to examine in depth the lives of ex-slaves during and after the process of emancipation that began in 1870. Studies by U.S. scholars have also drawn on new sources in archives located outside of Cuba and have stimulated discussions between Cuban historians and scholars from elsewhere who are interested in the history of the island.

Methodology and Sources

Most Cuban scholars have largely overlooked all forms of slave resistance but revolts and marronage, as they have paid scarce attention to the tools and methods of control that the colonial authorities and slave owners used. To understand the real motives behind the various forms of slave resistance, it is necessary to examine the different parameters and particular circumstances that occasionally led slaves to shift from using nonviolent forms of resistance to violent ones and vice versa. My analysis of the various forms of slave resistance is organized in categories that relate to their own meanings. Each of these categories is thoroughly examined in the six chapters that comprise the main body of this book.

In my investigation of some understudied themes in the historiography of slave resistance, I have mostly relied on primary sources. I have divided the forms of resistance into two main categories: violent and nonviolent. This division mirrors other scholars' categorizations of slave resistance.[35] In the category of violent forms of resistance, I include marronage, suicides, homicides, conspiracies, and revolts.[36] Nonviolent, or disguised, forms of resistance include slaves' use of the law, as well as cultural practices, such as music, dance, religious practice, gossip, folktales, and jokes. For each of these forms, I have gathered a significant number of documents from a range of different sources. Most of the documents are located in Cuban archives; however, I have also collected information from the National Archives (formerly the Public Record Office) in the United Kingdom, where I drew on letters of British officials and representatives in Cuba reporting on the relations between masters and slaves, and from the Archivo General del Indias in Seville, Spain.[37]

Local and central authorities examined rebellious and submissive slaves, free people of color, slave owners, and other witnesses, accused and accusers who people the pages of this study. These examinations constitute the bulk of the existing historical evidence regarding slave resistance in nineteenth-century Cuba. Therefore, their interpretation was one of the principal challenges of this investigation. The analysis of such testimony must come to terms with the omissions and manipulations that are part of any interrogation process. Even if one operates under the assumption that those who were under interrogation related events accurately to officials and that secretaries recorded their testimony precisely, hurdles to interpretation remain. Under the pressure of interrogation, individuals tend to misrepresent the truth or to lie outright for their own benefit. Others just refuse to talk. Silences thus need to be identified and interpreted when reading the testimony of any examined person. To borrow Winthrop D. Jordan's words, questioned people—especially those facing accusations of any type—and prosecutors are almost constantly engaged in a game of "cerebral calculations," in which each tries to get the best of the other.[38]

The testimony of African-born slaves, particularly those who were not yet "seasoned," may be less permeated by external influences than the testimony of other persons. Sometimes, especially in the wake of African-led revolts, interviewed slaves told very similar stories about their motives for rebelling.[39] There is no reason to doubt, after all, that physical punishment, frequent underfeeding, imprisonment behind the walls of barracks, and ultimately the very condition of being a slave were enough to prompt acts of resistance. At other times, however, particularly when free people of color and Creole slaves were involved in acts of resistance, slaves expressed their motives in a more complicated way, blending the aforementioned reasons with others that affected their lives.[40]

A second interpretive problem relates to the integrity and capacity of the prosecutors and secretaries who were charged with procuring testimony.[41] If the officials in charge of a given case forged declarations or resorted to coercion to obtain the results they wanted, testimonies were likely to be compromised. Moreover, if the secretaries' skills were limited or they did not wish to write for hours on end, they also were likely to manipulate the record in ways that could affect interpretation. Although declarations were supposed to be transcribed word for word, their manipulation is clearly visible in the record at times, as when secretaries used conventional formulas to transcribe the responses of the person being interrogated.[42]

Beyond these issues, more subtle problems remain. Our ignorance of the vocal inflections and physical gestures of the interrogated foreigners, slaves, slave owners, and neighbors are all substantial considerations when trying to interpret this body of historical evidence. Moreover, foreigners' testimony frequently had to be translated, leading to further problems of interpretation. On this matter, I endorse the opinion of Emilia Viotti da Costa: "The voice of the slave that reaches us in inadequate translations, through layers of biases and misperceptions, is barely audible."[43] This is not to say, however, that slaves did not speak, and that their—badly—recorded words mean nothing today.[44]

All these problems of interpretation aside, the criminal proceedings and the correspondence generated by them are suitable and useful materials with which to take a microhistorical approach to the study of slave resistance. Such an approach allows us to reduce the scale of study and to investigate the structure of the lives of common people—in this case, slaves, masters, and other witnesses. The documents related to the African slave experience on Cuban plantations are the kind of source that any historian would wish to find in the archives. Bearing in mind Carlo Ginzburg and Carlo Poni's statement that "if the sources are silent about or systematically distort the social reality of the lower classes, then a truly exceptional . . . document can be much more revealing than a thousand stereotypical documents," I decided to tackle the sources with the intention of finding the hidden "golden nuggets," the records of human experiences hitherto unknown to us.[45] What Ginzburg and Poni proposed was the intensive study of unique documents in order to recapture the interactions between elite and popular cultures. In tune with their microhistorical approach, I have tried to identify the historical protagonists involved in instances of slave resistance. Who were they? Where did they came from? What were their personal experiences? These questions have been my reference points in assessing their actions. Ultimately, it is the central question of microhistory that has led me to study forms of slave resistance: what can we know about peoples lost to history?

The present study, then, is mainly based on court records. These records show, sometimes in the slaves' own statements, how Cuban slaves saw their world and why they acted against their oppressors. Slaves were brought to court for a wide range of reasons. They were frequently accused of drumming and dancing, practicing "witchcraft," committing robberies, and engaging in illegal transactions. After suspicious fires, suicides, revolts, escapes, and strikes, authorities began judicial inquiries. Thanks to the Spanish colonial bureaucracy, we have a large number of documents produced by the prosecutors in charge of these cases. Although public prosecutors and their assistants usually manipulated declarations, slaves' words were sometimes copied literally, including their accents and grammatical mistakes.[46] Less frequently, in dealing with non-Christian slaves, prosecutors compelled the slaves to swear by their own gods or to offer their African birth names. Although the Cuban court records were generated by the authorities, they may be read against the grain, and that is precisely what I have tried to do here in a systematic way. In their statements, slaves revealed some aspects of their world to outsiders. These statements constitute the most reliable kind of source, enabling us to decipher the subjectivities that characterized the slaves' lives, including their intentions in regard to resistance. If these documents are properly examined, they can allow us accurately to reconstruct "meanings within their original context."[47]

Letters and diaries have also been useful sources. In both official and private correspondence, I found different opinions about virtually every one of the topics included in this study. Of special interest, due to their outsiders' perspectives, are letters written by travelers and foreign consuls about the situation in Cuba. Moreover, because they were not expected to become public, private letters sometimes offer acute critiques and impressions of the social tensions of the period, in which slavery was a major concern. Personal diaries have a tremendous importance as sources to the study of rebellions and especially marronage, since slave hunters (rancheadores) used to make daily notes of their activities. Finally, I have carefully analyzed newspapers in order to obtain a broader picture of Cuban society and plantation life.

In this study, then, I look at some particularities of slaves' daily life on nineteenth-century Cuban plantations. I do not intend to make the final statement about what was actually going on inside and outside the plantations. After all, this is merely a historical work, based on historical sources produced by officers belonging to the dominant colonial system and by planters and their acolytes. I hope, however, to offer a bird's-eye view of the green Cuban countryside, densely populated by African slaves and their descendants. While truthfulness may be a tricky and deceptive goal, I have not renounced it: I have striven to be as accurate as possible in examining and interpreting my sources. In this respect, I have observed the wise advice given by Carlo Ginzburg: "The obstacles interfering with research in the form of lacunae or misrepresentations in the sources must become part of the account."[48]

I

❧

The African Background of Cuban Slaves

Throughout the first half of the nineteenth century, sugar and coffee plantations in western Cuba were continuously supplied with new African workers. Although the Creole slave population increased steadily from the 1820s
onward, African-born slaves constituted the majority of the slave population
on these estates until well into the century. Several authors have looked at
the slave trade to Cuba, the conditions of slave transport, and the prices of
slaves.[1] Yet the slaves' cultural backgrounds frequently remain unexplored,
and this omission distorts our understanding of slave resistance. In this
chapter, I examine the principal cultural and historical backgrounds of
the slaves who landed in Cuba between 1790 and 1845, as well as the stereo-
types that circulated about the various African ethnic groups in nineteenth-
century Cuba. It is no easy task to sort out who these men and women were
and where they came from, but because they are the protagonists of this
study, it is an indispensable one.

After 1792, and especially after 1803, African slaves landed on Cuban shores
in ever greater numbers.[2] Following the abolition of the legal slave trade
within the British Empire in 1807 and among the remaining slave-trading
nations between 1820 and 1831, two Latin American territories—Cuba and
Brazil—benefited from the illegal slave trade. Cuban, Spanish, Brazilian, and
Portuguese traders challenged the Courts of Mixed Commissions and the

British navy in their attempts to enforce the end of the slave trade. Despite the risk of being captured, dispossessed, and sent to jail, these men carried on with their trade, and Cuban plantations continued to be well-supplied with cheap labor until the second half of the century.

It is well known that, as Louis A. Perez has recently stated, "Cuban planters early developed cultural profiles of the African *naciones*, associating dispositions and behaviors with specific ethnic groups."[3] These profiles were often based on the observations made by generations of slave owners and traders and were likely to be biased to some degree. But precisely because these profiles were based on observation—and because the goal of this cultural profiling was to amass wealth for those who benefited from slave labor—they also had some truth in them. Ethnicity was, without doubt, crucial for the planters. For decades, historians have argued that plantation owners avoided buying slaves from the same "nation" in order to avoid rebellions and collective suicides. This theory has become one of the classic myths among historians of slavery, who rarely argue against it. In her last book, however, Gwendolyn Midlo Hall has finally challenged this idea, basing her argument on the fact—shown in some cases analyzed in this study—that slave owners frequently preferred to buy new slaves from the same ethnic groups as slaves already living on their estates, because of the simple and often overlooked fact that in this manner they would find comfort in their seasoned compatriots and their own processes of settlement thus would be smoother.[4]

In Cuba, it was customary to give slaves a Spanish familiar name and, in place of a surname, a designation that referred to the slaves' tribe, ethnic group, language group, or geographical origin. In historical records, slaves are referred to by their Cuban names rather than their birth names, which in some cases provide scholars with information about where in Africa they came from. But there are also slave designations that simply muddy the waters for researchers. The term "Yoruba," for example, commonly used today to refer to peoples who live in the present republics of Nigeria, Benin, and Togo, "originated with European linguistic studies in the nineteenth century, and is strictly both anachronistic . . . and alien to the region."[5] Other problematic terms conflate West African coastal regions with ethnicity, due to the fact that the Spanish more often than not tended to place most of African slaves within broad nations, presumably to facilitate their marketing and sales. Thus, it is highly likely that under the term "Carabalí," for example, Spanish authorities frequently included most of the ethnic groups exported from ports of the Bight of Biafra, such as the Ibibios or the Igbos. The denomination "Mina" was also likely to serve as a broad definition for all slaves—such as the Ashantis or Fantees—who were exported from the Gold Coast and more specifically from the castles of Cape Coast and El Mina.

African ethnicities—and the confusion of scholars who attempt to disentangle them—have been the subject of recent studies. The work of Gwendolyn Midlo Hall, which challenges historians to rethink their understanding of African denominations used throughout the three and a half centuries of transatlantic slave trade, is particularly significant in this respect.[6] In the Cuban case, authorities, slave owners, travelers, foreign residents, and virtually everybody, including slaves, had preconceived ideas of the particular characteristics of every African ethnic group or "nation." Although I intend to avoid endorsing such preconceptions, there are reasons to believe that they were often well-founded and therefore worthy of being examined and interpreted.[7]

Throughout this study, one topic remains in the mist: the Islamic background of the slaves brought to Cuba. While African autochthonous religious beliefs have been at the core of several studies about the history of slavery in Cuba, Islam is rarely mentioned. There are no very satisfactory reasons why Islam in Bahia, for example, has been clearly identified and studied while in Cuba it has not. One obvious factor is that Cuban sources barely make any mention of slaves' Islamic background. But this absence of sources is in itself strange, and even more so if we again compare Cuba and Bahia—two regions with strikingly similar imports of African slaves in regard to ethnic and cultural backgrounds, particularly throughout the first half of the nineteenth century.

Many of the African slaves who arrived on both Bahian or Cuban shores after the turn of the century became enslaved in the interior of the Bight of Benin as a result of the jihad that Shuhn Uthman dan Fodio began in 1804.[8] The expansion of the Sokoto Caliphate and the collapse of the Oyo Empire were responsible for a sharp increase in exports through the ports of the Bight of Benin in this period. In the Bahian case, travelers, authorities, and even African slaves left a vast array of proofs of their Islamic beliefs. In Cuba, very little, if anything, was recorded.

How to identify Islamic beliefs among Cuban slaves is, then, a rather complicated undertaking. Although it is undeniable that Cuba received slaves from Islamized West African regions, such as the central Sudan or the Mandinga lands, the few testimonies we have—or that we think we have—about their presence are by no means conclusive. Some refer to Lucumí slaves—many of whom were likely the result of the expansion of the Sokoto Caliphate—who never drank alcohol. Other evidence is the product of purely speculative readings and interpretations of African names as they were re-corded by the Spanish authorities, such as those of some rebel slaves who participated in the rebellion of 1833 on the cafetal El Salvador. Some of their names seem to have been proper Muslim names, such as Lalani (Lawani), Achumo (Asunmo or Ismail), and Alu (Aliyu).[9] In the course of my research, I came across another possible Muslim slave, who was referred to as a Mand-inga Moro by the Spanish authorities—possibly as a way to define within one term both his ethnic origin and his religious orientation.[10]

The abolition of the transatlantic slave trade coincided with the fall of the Oyo Empire, in what is now southwestern Nigeria. As a result, there was an increase in the trade in former subjects of Oyo, who were known as Nagôs in Brazil and Lucumís in Cuba. In both countries, they were respected and re-nowned for their rebellious temperament.[11] The Lucumí slaves transported to Cuba were well-versed in the art of war. Before the arrival of the Euro-peans, they relied mostly on swords, spears, and bows and arrows. By the mid-eighteenth century, however, when the Empire of Oyo was at the height of its splendor, they had learned to use firearms and horses and had honed their military tactics to perfection. Cavalry played a remarkable role in the rise of the Oyo Empire; horses were used in most of the military campaigns initiated by the Oyo in the eighteenth century. In some cases, as in the inva-sion of Dahomey in the 1720s, the Oyo army consisted "entirely of cavalry," and the Oyo cavalry remained "significant" as late as 1823.[12]

The Oyo Empire had a long history of political alliances and coups d'état. According to Robin Law, the greatest territorial extent of the Oyo was achieved after one of these coups, when the Alafin Abiodun seized power from Basorun Gaha in 1774.[13] Coincidentally, the beginning of the end of Oyo dominance followed another coup d'état, this one carried out against Awole, the successor of Abiodun, in 1796. The "ephemeral greatness" of the empire came close to a terminal collapse in 1817, when the city of Ilorin and the Fulanis—a Muslim pastoral group from the north—allied against the Oyo.[14] From that moment onward, the slave ports of the Bight of Benin, which had formerly traded mostly in subjects from neighboring kingdoms like Nupe, Dahomey, and Borgu, began to increase their exports of Lucumís. The Oyo Empire finally crumbled in 1836 under the pressure of continuous attacks from the north, and in consequence more Oyo subjects were embarked to the New World—mostly to Cuba and Brazil—after that date.[15]

Daily life for the subjects of the Oyo Empire was infused with mysticism, which included a belief in life after death. This mysticism was a particularly important religious characteristic of most of the West African slaves who arrived in Cuba in the eighteenth and nineteenth centuries. The pantheistic religion of the Lucumí—known as the Santería in Cuba—was probably one of the richest in Africa. It had an oracle for divination of the future, shrines where gods and ancestors were worshiped and called upon, masquerade ceremonies where the worlds of the living and the dead mixed with each other, and, notably, a "very firm belief" in the reincarnation of souls.[16] To the subjects of the Oyo, the dead were "simply removed temporarily to another sphere."[17] To contact them, the Lucumí would celebrate masquerades in which they would wear their Egungun costumes. The role of clothes among Oyo subjects should not be underestimated. The proper clothes were essential when going to war, but they were also crucial when the Oyo were contacting their gods and ancestors. According to P. S. O. Aremu, clothes were the "magnetic forces" that drew the living and the dead together.[18]

Although the masquerade and the Ifá divination system were central to Lucumí beliefs, worshiping at sacred shrines also constituted an important part of the Lucumí's cosmological dynamics. Every person was protected by a specific Orisha, and every Orisha had his or her own sacred shrine. Orishas had individual personalities and specific characteristics that defined them. Colors and attributes, for example, separated Yemayá from Oshún, and Oggún from Shangó. Shrines were necessary, a circumstance that might well explain—at least in part—the process of transculturation by which Orishas were worshiped in Catholic churches in various parts of the New World in the guise of Christian saints and virgins.

A similar tale can be told about the Ararás, known in some Caribbean territories as the Aradas, who also hailed from West Africa. Most of the Ararás sent to the Americas belonged to an Ewe-Fon-speaking people who had settled in the former kingdom of Dahomey, which now includes the vast majority of modern Benin and Togo and a part of Ghana.[19] Beginning in the seventeenth century, Araráse were embarked from the most notorious slave ports of the Bight of Benin, namely Whydah, Porto-Novo, and Cotonou, among others. Their religion—or, to be precise, the version of it that developed in the New World—is perhaps the most enduring and popular of all the belief systems brought to the New World by West African peoples. In Dahomey, Voodoo was an established religion with a theological system of ritual worship. The very meaning of the term "Voodoo"—to draw water—reflects the cycle of life: birth, death, and rebirth. Ever since Voodoo played a crucial role in the Haitian Revolution, it has been widely studied by scholars. In Voodoo ceremonies, dancing, singing, drumming, and especially costumes all are extremely important. In addition to spiritual entities called Loas, the Araráse believed in the perpetual existence of the souls of the dead. After people died, their souls remained on earth, offering guidance and help to their relatives.[20]

During the eighteenth century, Dahomey became one of the biggest and most powerful kingdoms in West Africa. According to Elisée Soumonni, it raided the neighboring kingdom-cities of Ketu, Sabe, Idaisa, and Mahi in search of slaves "almost annually" throughout the eighteenth century.[21] Dahomey's participation in the slave trade intensified after 1818, when the fall of the neighboring Oyo Empire and the ascension of King Ghezo in Dahomey combined in the kingdom's favor.[22]

The military experience of the subjects of Dahomey was exceptional. Araráse played a leading role in the great slave revolution of Saint Domingue and in many slave revolts that occurred in nineteenth-century Cuba.[23] As early as 1690, the kingdom of Whydah was said to be importing a thousand firearms a year. Not long afterwards, Pieter de Marees wrote about the organization of the Whydah armies, mentioning platoons that were subdivided into smaller military units, each led by a flag bearer and a commander holding a large umbrella.[24] By the mid-eighteenth century, Dahomey not only had a respectable and well-organized army but also a fleet of large canoes armed with "small iron cannons."[25]

Many of the slaves who arrived on Cuban shores between 1790 and 1870 began their journeys in the Bight of Biafra, in southeastern Nigeria, from the ports of Bonny and Old and New Calabar. Although these men and women

belonged to different peoples and ethnic groups, they were known in Cuba
as Carabalís, after the Calabar River. Carabalí slaves were reputed throughout
the New World to be fierce and prone to commit suicide.[26] In Cuba, they often led or actively participated in slave conspiracies and revolts. Considering
the history of the peoples who were included under this name, such a fact
hardly seems surprising. According to Douglas Chambers, approximately
80 percent of the people sold as slaves from the ports of the Bight of Biafra
were Igbo-speaking.[27] Igbos were known to be an extremely proud people
with a high regard for independence and autonomy. They enjoyed political
liberties within a system of village democracy that contributed to the decen-
tralization of power. Igbos were also regarded as a deeply religious people,
with strong beliefs in the influence of their ancestors and the reincarnation
of their souls. The Igbos and their neighbors the Ibibios had "similar cultural patterns, customs, and traditions."[28] They respected and feared the Juju
deities and celebrated the New Yam Feast in those deities' honor.[29] Igbos had
a deep knowledge of poisons and used them in ordeals. In their religious
and secular festivals, clothing played a significant role.[30] Although they did
not witness or participate in large wars, they frequently fought among themselves; these were small but intense local battles in which women fought
alongside men.[31] In Cuba, the Efik, neighbors of the Igbos and also labeled as
"Carabalí," formed a secret society known as Abakuá that has preserved many
elements of the culture and beliefs of this group from the Niger River delta.[32]

Also heavily decimated by the slave trade were the people of the Gold
Coast known in Spanish America as Minas. The term "Mina" derives from
the slave factory of São Jorge de El Mina, which was first Portuguese and
later Dutch and was located west of Accra, the capital of what is now the
Republic of Ghana. There were several slave factories along the shore of the
Gold Coast. Not far from El Mina castle was the Cape Coast castle, where
thousands of slaves began their Middle Passage in British ships. Many other
factories were established along this coast by the Dutch, the Danish, the
Swedish, the British, the Portuguese, and even by slave traders from Bran-
denburg. Slaves from the interior of the Gold Coast were also traded to Hispanic America largely under the generic name of "Mina." They came from
a very unstable region, over which slave traders and British soldiers were in
a constant state of alarm even in the nineteenth century due to the military
strength of the Ashantee Empire.[33]

Throughout the Americas, Gold Coast slaves were considered to be prone to rebel. They were central figures in several New World slave revolts of the eighteenth and nineteenth centuries, including the Maroon Wars in Jamaica and the great slave revolt of Demerara in 1823.[34] The Ashantees and Fantes of the Gold Coast were familiar with warfare and firearms at least from the late fifteenth century. In the early seventeenth century, de Marees wrote about the firearms trade between the Portuguese and the "Elminas"—most likely the Fantes—and stated that the Dutch were not only selling guns but also teaching the local people how to use them.[35] According to Ray A. Kea, musketeers had become a common part of Gold Coast armies by the seventeenth century.[36] The Elminas were also "the fittest and most experienced men to manage and paddle the canoes over the bars and breakings," which helped them to empower their states.[37] The Ashantees, in contrast, became "one of the most military powerful and structurally articulate polities in all of West and West-Central Africa, rivaling the earlier savannah states in complexity if not territorial expanse."[38] The centralized government of the Ashantees and their eventual success in conquering all the neighboring states created what Michael Gomez has called "an atmosphere of insecurity" throughout the region. This atmosphere was characterized by frequent imperialist wars directed from the Ashantee capital, Kumasi.[39]

Akan beliefs in life after death had their "very origin" in the word the Akan used for soul, "Kra" or "Okra."[40] As Sam Akesson notes, for the Akan the "souls of new born children are either emanations of ancestral souls or reincarnated former lives."[41] For the people of the Gold Coast, the reincarnation of souls was central to their lives. This belief—common among most of the West African peoples—was not a "vague hope for life after death," but a certainty.[42] For the Akan, death meant a return, sooner or later, to their family or group. As Akesson remarks, "Their beliefs in the immortality of the soul ... explain a major factor in Akan life; it is the concept which explains the perpetuity of the clan, the tribe. Individuals may die, but that does not affect the group."[43]

Another important center for the shipment of slaves to the Americas, and specifically to Cuba, was the region of Sierra Leone and Liberia. The slaves sold in the factories of the rivers Nuñez, Pongas, Grande, and Gallinas were known generically in Cuba as Mandingas and Gangás. The Mandingas were "a subgroup of Mande speakers" and were often referred to as being Muslim. John Matthews, for example, recalled that "Mandingoes" were "Mahomedans" and "[as] zealous promoters of their religion as even Mohamed himself could wish."[44] The Mandingas seem to have been involved in the religious wars that plagued this region from the 1720s, when a holy war, or jihad, was launched by the allied Fulbe and Jallonke Muslims toward the peoples living in the Atlantic regions.[45] According to Walter Rodney, from the mid-eighteenth century, slaves exported to the New World from the ports of Sierra Leone and Liberia were mostly from the interior.[46] Mandingas, then, were probably used to fight wars to defend themselves, for which they developed their own military techniques. Willem Bosman mentioned that the people of Sierra Leone and Guinea poisoned their arrows and spears as early as the seventeenth century.[47] The seaside inhabitants of these regions also fought their wars at sea. A sixteenth-century Portuguese traveler wrote about the "huge canoes of the Sierra Leone River" that carried warriors "armed with spears, shields, and arrows."[48] There are also references to these peoples' use of the horse from the early eighteenth century.[49] Their trading skills and the fact that they were among the first Islamized groups in West Africa transformed their language, Mande, into a sort of commercial and political lingua franca in the area.[50]

Since Mande was an "extremely homogenous" language, it seems reasonable to conclude that many of the slaves who entered Cuba under the name of "Gangás" also belonged to the Mande family.[51] There is no evidence to suggest that the Gangás constituted a singular ethnic group or subgroup. The most feasible explanation for their name—which has been a headache for scholars over the years—is that "Gangá" was derived from "Gbangba," the name of a river close to the border between the present-day republics of Sierra Leone and Liberia. This supposition is reinforced by the fact that many of the Gangá slaves' subdenominations related in one way or another to the local geography and culture of the hinterlands of present-day Sierra Leone and Liberia.[52] The Gangá Kissi—renowned for their manufacture of clothes—were almost certainly from the Kisigudu area, while the Gangá Fay received their name from the neighboring vicinity of Vai or Vey, along the Gallinas River.[53] The Gangá Cono were most likely from the Konno area, north of Gallinas, while the Gangá Mani were probably from the area surrounding the Mani River, which is located on the border between Guinea and Liberia.[54] A second theory about the name "Gangá" was proposed decades ago by G. Aguirre Beltrán, who suggested that it derived from the term "Gangara," which was used by the Arabs to denominate the Mande.[55]

Like the Mandingas, those groups that arrived in Cuba under the name of Gangás were familiar with warfare. They were also targeted by the Fulbe-Jallonke jihad of the 1720s and, as a result, were likely to constitute one of the main sources of slaves in the area. With this fact in mind, it seems reasonable to suggest that a large number of these Gangás were not Muslims, a fact that possibly played a central role in their enslavement.[56] In Cuba, the Gangás were sometimes considered to be noble and good-hearted, but they were also labeled as thieves and were thought to be skillful at escaping.[57] They were often leaders in plots and revolts and willing participants in virtually every action of slave resistance, from sabotage to marronage. Although their precise origin and their cultural legacy are still matters of debate, there is little room to doubt their impact in eighteenth- and nineteenth-century Cuba, on the plantations in particular.[58]

The Gangá—or the Gallinas people, as they were called by West African travelers—were reputed for having "fine towns like the Mandingoes."[59] J. M. Harris considered them "inveterate gamblers," and H. C. Creswick described their character as "eminently light and buoyant."[60] Creswick also made reference to their habit of dancing and playing "all night long" and to the way in which they used their tamanden—war drums—to communicate messages from town to town.[61] The Gallinas believed in the power of oracles and in the magical properties of amulets. In 1894, Scott Elliot noticed that "the wearing of amulets on the arm or neck" was common among the Gallinas, who believed that the amulets would keep them free "of devils."[62] Creswick also mentioned the amulets—called gergrees—which, in his words, were "an institution" among them.[63] According to Creswick, in wartime the important leaders could be seen "almost loaded with charms of different kinds and degrees of virtues."[64]

No doubt the largest number of slaves who arrived in Cuba began their transatlantic journey in West-Central Africa. These slaves were known in Cuba as "Congos."[65] Congo slaves were embarked in the vast region between the ports of Loango and Benguela, which fall along the coastline of present-day Gabon, the Republic of Congo, the Democratic Republic of Congo, and Angola.[66] Congo slaves were recognized in both Cuba and Brazil as being very brave but at the same time as noble, hardworking, and servile.[67] They played important roles in the Stono Rebellion in 1739, in the great revolution of Saint Domingue, and in various northeastern Brazilian quilombos, most notably in Palmares.[68]

In terms of their experience in warfare, Congo slaves were no less experienced than other slaves introduced into Cuba. Since the fifteenth century, the kingdom of Kongo had been respected and feared even by Europeans. John K. Thornton states that according to various European visitors to the region, "Angolan armies seem to have been professional, at least in the late sixteenth century."[69] They possessed "facility in hand-to-hand combat with sword, club, battle axe, and stabbing spear, and in some cases use of the shield."[70] The slave trade in this region seems to have been especially harsh and closely related to the phenomenon of "escalating warfare over time."[71]

A last significant group of slaves to arrive in Cuba were embarked in the East African ports of Quilimane and Mozambique. Although these men and women had different ethnic backgrounds, they were often known as "Macuás" or "Mozambiques." Like the Congos, Macuás belonged to the Bantu family. Since these peoples, especially those who lived alongside the Zambezi River, were in permanent contact with their western neighbors, it is feasible to infer that they were also subjected to the wars that had plagued West-Central Africa from the sixteenth century. Macuá or Mozambique slaves made up a considerable portion of those carried to Cuba in the transatlantic slave trade.[72] As late as the 1820s, the Portuguese were still restricted to the coastal areas of Mozambique and "were not allowed to enter the Makua and Yao territories."[73] On Cuba, the Macuás were reputed to be rebellious and disaffected. Some of them, like Domingo Macuá, became maroon leaders of almost mythological stature, while others participated in various types of collective resistance. In 1819, a Portuguese traveler reported that the Macuás "were demons."[74] Their culture and religious beliefs, although inevitably transformed over time, still remain very much alive in some places in the western part of the island, including the capital, Havana.[75]

As this brief survey has shown, the vast majority of African men and women who were forced to endure the transatlantic voyage to the Americas were familiar with war. War was an integral part of their lives and in many cases was the reason for their enslavement.[76] From the sixteenth century onward, Europeans, by means of their military techniques and technologies, steadily penetrated West and West-Central African territories. Africans who arrived in the New World—at least, most of those who arrived after 1790—thus had some knowledge of European weaponry and military techniques as well as of their own traditional ones. Many among them knew about modern military formations, guns, cannons, and so on, but they also had experience using spears, bows, arrows, and clubs. Their enslavement was the product of a time of change and expansion in international commercial networks. When they were traded to the Americas to work as slaves, they carried with them their life experiences, their religious beliefs, their knowledge of the world, and, perhaps most importantly, their knowledge of war, which made many among them into time bombs.

2

Homicides, Conspiracies, and Revolts

The crowd may riot because it is hungry or fears to be so, because it has some deep social grievance, because it seeks an immediate reform or the millennium, or because it wants to destroy an enemy or acclaim a "hero"; but it is seldom for any single of these reasons alone.

—George Rudé, The Crowd in History

On July 1, 1840, Tranquilino, a slave from the coffee plantation Empresa, killed his mayoral in a quarrel. Moments later, more than fifty of his companions headed to the house of Empresa's owner, José Cantor Valdespino, and assassinated him. Shortly before these events unfolded, Valdespino had sent one of his most loyal servants, Julián, to the nearby village of Ceiba del Agua in a desperate attempt to get reinforcements who could help him survive the tumult. Julián ran as fast as he could to the office of Captain Sixto Morejón, but his effort was futile. When the militia led by Morejón arrived in the plantation's batey, Valdespino was already dead, and the rebellious slaves, most of them Lucumís, were hiding behind a pile of stones or escaping across the coffee fields. Seeing the resolution of the rebels, who, "obeying the order of their captain," attacked his men with stones, Captain Morejón decided to open fire regardless of the consequences.[1] At the end of the day, he counted the casualties and regretted the loss of both Valdespino and the mayoral but said nothing about the rebels except that almost all of them were of the Lucumí nation.[2]

The reason for the two homicides and the revolt was apparently the bad treatment to which the mayoral had subjected the slaves under his command.

Tranquilino, whose name, curiously, means "tranquil" in Spanish, decided to take the mayoral's life after being whipped for refusing to work. While fixing the roof of a house a few days before the revolt, Tranquilino had fallen to the ground and hurt himself very badly. The mayoral, who had witnessed the accident, did not consider the injury severe enough to make allowances for, and he forced Tranquilino to work as usual. The day of the incident, the mayoral tested Tranquilino's patience by trying to flog him, thus triggering the tragedy. Noting Tranquilino's resolve moments before the attack, Valdespino intervened in a last attempt to change the slave's mind, asking him why he did not look for a padrino to represent him, since he was so upset. Tranquilino's answer was defiant. He replied, "I did not do it because the only padrino is God, because if I get a padrino just for today, tomorrow you will not defend me [against the mayoral]."[3]

These two murders and the subsequent revolt highlight the main issues I will examine in this chapter. How did homicides lead to revolts, and how did revolts provoke more homicides? To what extent did ethnicity, language, leadership, and previously acquired knowledge of warfare affect these events? What were the differences between movements led by slaves and those led by free men?

Academic studies of slave revolts in Cuba have predominantly focused on the emergence of conspiracies and revolts of a revolutionary type, such as the ones described above.[4] As a result, other movements, including those organized and carried out by African-born slaves, have been systematically underestimated or ignored, as have some more common types of violent resistance that can be thought of as African rather than as politically motivated or revolutionary. Homicides and suicides, both usually linked to African-born slaves, have not attracted much attention in most of the recent works on slave resistance in the Americas. In this chapter I look at the primary conspiracies and revolts that occurred in early nineteenth-century Cuba, intentionally avoiding a rigid chronological scheme and prioritizing instead the characteristics, contents, and context of each event and its place in history.

Slave Participation in Abolitionist Movements

When Captain-General Jerónimo Valdés wrote to the Spanish minister of overseas affairs in September 1843 that the island "was still inalterably quiet," he was well aware that his words were entirely false.[5] Beginning the moment Valdés had taken control of the Cuban government two years earlier, a forceful, fast, and extremely dangerous succession of slave movements had shaken the western part of the island, threatening the very basis of the Spanish colonial system. The most remarkable aspects of Valdés's years in office—and also of the previous decades—were the African-led slave revolts that were planned and carried out largely in plantation areas that were not far from the colonial capital, Havana. Movements led by free men (mostly Creole) also constituted a serious problem for Cuban authorities.

Many of these movements were premeditated agreements among slaves to bring an end to their servitude in the sugar mills and on the coffee plantations. Others were the barely organized attempts of the desperate to escape the unremitting routine of hard work and daily punishments, beginning in isolated and unplanned acts of resistance against the injustices of owners, overseers, slave drivers, and others.[6] Taken together, these rebellions were a panoply of violent acts of resistance that deeply preoccupied every captain-general who governed Cuba in the first half of the nineteenth century. Valdés was well aware of the magnitude of the threat. From the moment of his arrival on the island until his departure from it, he was informed of at least twelve important incidents of slave rebellion. This fact strongly contradicted his letter to the Spanish minister of overseas affairs.

In December 1844, barely a month after Valdés ceded his post to his successor, Captain-General Leopoldo O'Donnell, a vast conspiracy was uncovered, involving slaves who were organized and led by free colored men and who were apparently supported by the British. This very well-coordinated movement had active cells from Havana to Santa Clara, with Matanzas, Cárdenas, and the surrounding countryside as its main centers. In Cuba, its name, La Escalera, has been synonymous with repression and racial hate practically from the moment when its leaders began to be executed behind the walls of the fortresses of Havana, Matanzas, and Cárdenas.

La Escalera is undoubtedly the best-known and best-documented Cuban slave conspiracy. The trials of the plotters fill more than fifty bundles of documents, today located in the Archivo Nacional de Cuba. Hundreds of additional documents are dispersed throughout libraries and archives in Cuba, Europe, and the United States. The fact that the most comprehensive book written on the conspiracy did not use any information from Cuban archives suggests just how disseminated and abundant this documentation is. La Escalera signaled a change for Cuban slaves: the year 1844 was at the same time the peak of whites' fears about slave rebellion and the end point of the longest cycle of slave revolts ever witnessed in the Americas. After La Escalera, slaves' involvement in Cuba's social movements fell off, returning to its previous levels only at the start of the Ten Years' War in 1868.

In the wake of several slave outbreaks that occurred around Matanzas and Havana between 1842 and 1843, Spanish authorities uncovered a plan to end the slave system in Cuba and establish a political regime similar to that of the neighboring Republic of Haiti. A long-lasting academic discussion has ensued about whether such a conspiracy in fact existed. Even after Robert L. Paquette and Rodolfo Sarracino made their excellent contributions to this debate, it remains alive. Today, we know that hundreds of free blacks and mulattos were conspiring against colonial rule.[7] It also seems to be true that they were receiving moral support from the British consular officials who lived in Cuba at the time. It is probable that the so-called leaders of the conspiracy were aware of the international events of their time and of the political ideas introduced by the French and Haitian revolutions.

As Paquette has rightly pointed out, La Escalera "has received uncommon attention in Cuba."[8] Topics of scholarly discussion have varied over time but have centered on whether the plot truly existed and the exact nature and level of involvement of some of the most notorious characters associated with it. Among the writers involved in these debates were some nineteenth-century intellectuals, including Manuel Sanguily, José de Jesús Márquez, Vidal Morales y Morales, and Joaquín Llaverías.[9] New generations of scholars have offered new judgments on a variety of matters.[10] Despite their divergent opinions, most of the scholars who have studied La Escalera agree about the importance of the foreign instigation and backing of the conspirators.

Indeed, several testimonies mention foreign influence and assistance. Miguel Flores, one of the most important ringleaders of the conspiracy, declared that British consuls David Turnbull and Joseph T. Crawford encouraged the slave conspirators to go ahead with their project.[11] Antonio Lucumí, a freedman from Gibacoa, declared that his mulatto friend José Antonio Ramos had visited three British ships and that the "English talked to him and told him that they would bring them guns, gunpowder, bullets, and sabers."[12] Basilio Pérez, a free black from Santa Clara, was even more specific when he declared in April 1844 that "the referred outbreak was the Englishmen's idea, and that they [the British] should arrange everything with the people from Santo Domingo in order to send weapons and a general to command the rebels once the planned uprising had begun."[13]

Even more remarkable is the fact that most of the alleged leaders of the conspiracy were well-educated mulattos from the cities of Havana and Matanzas. The supposed ringleader, Gabriel de la Concepción Valdés (Plácido), was a poet well-known across Hispanic America.[14] His comrades—all literate men—included musicians and artists. They read newspapers and discussed their ideas in cafés and at private meetings. They also frequently traveled and, in consequence, were able to speak more than one language. Until the repression of La Escalera began, they were widely accepted at the most select urban Cuban gatherings as "adapted" and "civilized" people of African ancestry— as an example to be followed by other Africans and African descendants who inhabited the island.

It remains difficult to determine whether these men were planning a large revolt in Cuba or whether they were masterminding several interconnected plots. It is clear, however, that they had something arranged. More important, they were well aware of the evils of the Cuban slave system and convinced that it was possible to bring it to an end.

The second largest Cuban movement of rebellion was the conspiracy of Aponte. Uncovered after a series of slave revolts early in 1812, this was probably the most dangerous conspiracy ever planned in Cuba. Following the revolutionary ideas of the era, some free mulattos and blacks—supported by a few slaves—plotted to terminate the slave system. One of the main ringleaders of the movement, the free black José Antonio Aponte, was a sculptor. Among his accomplices were literate artisans, cabildo masters, militia soldiers, and also a few urban slaves.[15]

The influence of the Haitian Revolution was significant in this conspiracy. The primary evidence of the conspiracy collected by the Spanish authorities consisted of a book of drawings found in the Aponte's house. This book—unfortunately lost today—contained drawings of George Washington and the black kings of Abyssinia. Even more disturbing to the authorities were images of the most important leaders of the Haitian Revolution, among them Toussaint L'Ouverture, Henri Christophe, Jean Jacques Dessalines, and Jean François.[16] When the plot was uncovered, there were at least three tiny drawings of the Haitian king Henri Christophe circulating in Havana among the conspirators. Simultaneously, Jean Barbier, a French black and an active member of Aponte's circle, was assuming the identity of the deceased Haitian general Jean François and inviting free colored people and slaves to join him in his fight against the Spanish slave system.[17]

Thanks to the research of Matt D. Childs, we are able to know the possible aims and motives of the 1812 plotters. The Aponte conspiracy was perhaps the broadest-based of all nineteenth-century movements of resistance. Despite being put in motion mainly by free blacks and mulattos of urban extraction, slaves also played a significant role in its development. The urban-based circle of Aponte's close friends and collaborators maintained a tight relationship with slaves from the countryside surrounding Havana. Its members often traveled to nearby plantations, encouraging the slaves to participate in a general uprising against their masters. Members of Aponte's circle also shared their ideas and knowledge about freedom. They participated in dances and the beating of drums in traditional African styles.[18] The combination of different cosmologies and cultural backgrounds is one of the most important characteristics of the Aponte conspiracy. At the same time the slaves—encouraged by the promises of their free colored friends— were planning to kill all whites and become owners of the land, free blacks and mulattos had their sights on changing the social system as a whole. Instead, the conspiracy was uncovered in March 1812, and its main protagonists were executed soon thereafter.

Just six years earlier, in 1806, again not far from Havana, another movement involving both free and enslaved men had been uncovered.[19] The plotters, most of them from the small village of Guara near the southern port of Batabanó, planned to take the capital of the island. To achieve their goal, they considered it necessary to talk to the "old blacks" who had labored years before in the construction of the fortress of La Cabaña, the city's main defense. The conspirators' tactics were not particularly sound, however: they planned to send their men over the roads, "killing people and stealing horses to ride, and taking all the weapons they could find."[20]

The conspiracy was arranged by two slaves, one of them a Creole, Francisco Fuertes, and the other a man from Saint Domingue, Estanislao. When interrogated, both men made similar declarations about the plan. The former French slave's experience and the Creole's ability to write and read combined to create a new threat to the peace and lives of the white residents of the region. Another significant peculiarity of this conspiracy was that its masterminds relied on slave drivers[21] to lead the slaves, since they "were the principals whom all the slaves [should] obey."[22]

During the 1790s, a number of conspiracies and revolts broke out that were somehow influenced by the news of continuous slave unrest in the neighboring French colony of Saint Domingue. As some historians have rightly pointed out, refugees from Saint Domingue and their slaves, as well as deported blacks, arrived in Cuba following the onset of the Haitian Revolution in 1791.[23] This fact led Cuban captain-general Luis de las Casas to ban the introduction of French-speaking slaves into Cuba as early as 1796.[24]

One revolt in which "French" slaves played a leading role occurred on the estate Cuatro Compañeros in July 1795. The ringleader of this movement was a slave known as José el Francés, or "José the Frenchman," who probably hailed from Saint Domingue.[25] Another revolt took place in Puerto Príncipe in June 1798. It involved urban slaves, who presumably had knowledge of international events, and again one of the leaders was known as "a French[man]."[26] Also in 1798, an extensive plot conceived by freemen and slaves working together was discovered in the city of Trinidad and its surrounding countryside. The name of one of the ringleaders, Josef Maria Curaçao, suggests his connection to the Dutch Caribbean island of Curaçao.[27]

Although all these plots and revolts have been regarded as important in Cuban history, they were exceptions rather than the rule. African-born slaves carried out most of the recorded acts of open insubordination, namely, revolts and the murder of their owners and other authority figures. Often they did so for reasons strongly related to their codes of behavior and beliefs and to their previous social and cultural experiences.

Taking Whites' Lives: Homicides and Their Lost Place in History

Homicides carried out by slaves have not received as much scholarly attention as one might expect. The assassination of masters and overseers was common in plantation Cuba throughout the period of slavery, yet historians consistently underestimate its importance, perhaps because the individualistic character of homicide does not fit in very well with the stereotype of the rebellious slave. Even today, slaves' homicides receive far less attention than they deserve. Gloria García has commendably drawn attention to them as a valid subject of study. She stresses the variety of circumstances that could

lead to their occurrence, as well as their important place in the different ap-
prehensions and fears of Cuban authorities and planters.[28]

Although most of the homicides recorded in colonial court records were
physically violent, it seems certain that slaves often found safer ways to get
rid of their masters and overseers. The account of the March 1844 poisoning
of Juan Bautista Després, owner of the sugar mill Unión, suggests one av-
enue many slaves might have taken. Several testimonies given by slaves from
Unión and from the neighboring estate, which was owned by Pedro Lorenzo,
affirmed that slaves Pedro Gangá and Jacobo Lucumí blended some poisonous powders into their master's coffee and that he died soon afterward.
Some days later, Pedro and Jacobo tried to poison the rest of Després's family, but when Jacobo served the lethal coffee, the family members all found
it unpleasant and refused to drink it.[29] A similar plot took place around the
same time in the neighboring paddock, which was owned by Juan José Marquetti. Slaves Lino and Canuto told the Military Commission that Camilo
Macuá tried to convince them to put poisonous powders in Marquetti's coffee. Camilo denied all the charges and eventually escaped alive.[30]

Homicides were often acts of retaliation. Floggings, insults, or vitupera-
tive language could drive slaves to dangerous and unexpected responses;
there were hundreds of such incidents in nineteenth-century Cuba, with
diverse results. Slaves did not always end up killing or injuring their masters.
Often they were brought under control or killed before they could achieve
their revenge. At other times, however, homicides triggered huge revolts.

Slaves might retaliate even for small reasons. In January 1824, Pedro José
Mandinga killed the manager of his plantation for calling him "lazy and
drunk."[31] In another case, José Delgado was murdered by his slaves Mariano
Gangá and Severino Lucumí following an argument. Both slaves declared
that they killed their master because he had tried to correct them. Curiously,
though, they blamed their behavior on a supernatural force: both agreed that
their master was generally a good man and that the devil had possessed their
bodies and minds at the time they murdered him.[32] A third case illustrates
how dangerous any new prohibition imposed on slaves could be for white
owners and overseers. In April 1844, Pedro Gangá recounted that he and his
comrade Jacobo Gangá had planned to kill the manager of the sugar mill
Unión in Guamutas after he had decided to lock them up every night, thus
preventing Jacobo from going to meet a black woman he liked.[33]

Not surprisingly, physical punishment most frequently inspired aveng-
ing actions. In September 1826, Luis Lucumí and Esteban Mina beat their
mayoral until he was unconscious. Both slaves had been whipped by the
mayoral for refusing to move the bodies of some dead pigs.[34] In another
instance, Gervasio Carabalí almost killed three white people outside the
small village of Bainoa, west of Havana, on the morning of April 22, 1833.
The night before, he had been in town carousing without the consent of his
owner. When he returned home the following afternoon, he was slammed
in the face, whipped, and bashed with a large cropping tool. Under such
provocation, he pulled out his knife and stabbed a number of people while
fleeing the premises. Gervasio died in prison three months later, the victim
of cholera morbus, while awaiting the verdict for his offense.[35]

When individual actions became collective thanks to the solidarity or
commitment of fellow slaves, single retaliatory actions could develop into
mass revolts. This frequently happened in the Cuban countryside throughout the first half of the nineteenth century.

An Extraordinary Sequence of Slave Rebellions, 1795–1844

In September 1827, with five other slaves, Rafael Gangá faced his master,
another planter from the vicinity, some white employees of the two planters, and a slave named Fermín who was loyal to Rafael's master. Just before
noon, Fermín threatened to release his dogs on the six disobedient slaves,
who had decided not to work. According to Rafael Gangá, his comrade Francisco approached Fermín and asked him, "Hey, you son of a bitch, why are
you going to release the dogs?" Fermín did not answer; instead, he set his
dogs on the slaves without hesitation. Moments later, all was confusion. The
slaves ran away; the two slave owners and their employees tried to capture
them. The outcome was a massacre. Five of the six slaves died in the fighting, and Rafael Gangá, the only survivor, perished more than two weeks later
as a result of his wounds. Both masters, Agustín Hernández and Domingo
Luis, lost their lives in the fight, as did some of their employees, including
Fermín (and his dogs).

After running from the dogs, Rafael and his comrades hid behind a
wooden fence and from there confronted not only their masters and their
employees but also reinforcements sent from the neighboring village of Catalina de Güines. To defend themselves, they made use of two carbines, some

bows, and more than two hundred arrows. According to Rafael, their only motive in fighting back was their loyalty to their captain Tomás, to whom they had been tied for military reasons since the time they had lived in Africa.[36] In this case, after taking the lives of those who were directly oppressing them, the slaves went into open revolt, causing more deaths and becoming a serious concern for whites in the vicinity of Catalina de Güines.

This revolt suggests the intriguing connection between homicides and slave rebellions in colonial Cuba. Somehow, they were often both the cause and the consequence of each other. Homicides often ended in open uprisings, and revolts caused homicides as they developed. The use of both European and African weapons also raises the question of what knowledge slaves had of different types of weaponry. Finally, the only survivor testified that the main reason for revolt was the slaves' great respect for their captain and their obligation to follow him to war. These three elements of this episode of homicide and rebellion were common to most of the African-led slave movements of the first half of the nineteenth century. The experiences of Rafael Gangá and his comrades were not extremely different from those of the thousands of other slaves who arrived on Cuban shores during this period. They had a profound sense of commitment and solidarity and certain knowledge of how to make war. Many of them while still in their native African lands were familiar with the weapons of the whites. And they showed their obedience to African leaders with a dignity that often inspired them to die fighting for their freedom.[37]

The 1798 rebels of Puerto Príncipe were mainly plantation slaves from the Calabar region of West-Central Africa.[38] Presumably, the revolts of 1817 and 1822, both of which took place on sugar plantations, were also of this type.[39] Unfortunately, these two slave movements are poorly documented due to the lack of surviving historical records.

The first recorded rebellion of the nineteenth century that was totally organized and commanded by African-born slaves took place in the valley of Guamacaro, near Matanzas, in June 1825. Between midnight and midday on June 15, more than two hundred slaves—lacking a solid plan but relying on their knowledge of magic and warfare—rebelled against their masters and overseers.[40] In the process, they disproved white Cubans' supposition that by mixing the different African ethnic groups they could prevent slave plots, on the grounds that tribal rivalries and the lack of a shared language among the slaves would prevent cooperation. The rebellion of 1825 in Guamacaro—a small village east of Matanzas—was conceived and organized by three slaves. Pablo Gangá, its mastermind, was a coach driver. Federico Carabalí was a respected sorcerer and a chief figure among the slaves, who believed in his supernatural powers. Lorenzo Lucumí, who took charge of the troops once the revolt began and who was assassinated ten days later after a magnificent fight against his hunters, was the third chief conspirator. It seems that Lorenzo was well-versed in warfare. He commanded the rebels wisely and was defeated only because his opponents' forces were better armed and outnumbered his men.

Lorenzo, Federico, and Pablo conspired for months to end their miserable situation. Although their reasons for rising against their masters were not clear, even to the authorities, it seems that they were not happy with their enslaved condition and believed that they could change their way of life by escaping from the plantation—but only after taking revenge upon their masters. A large majority of the rebels who joined the uprising were of Carabalí origin, although their main leader during the rebellion was a Lucumí. The revolt had long-lasting effects. Soon after the rebels were put down, the governor of Matanzas issued a local slave code to ensure his control over the extensive slave population of the region.[41] Five years later, a supposed plot was uncovered on the coffee plantation Arcadia in the same region of Guamacaro. When informed about the conspiracy, Captain-General Francisco Dionisio Vives immediately recalled the 1825 rebellion.[42] One of the slaves accused in 1830, an African recently arrived on the island named Bozen Mandinga-Moro, declared that "a long time ago . . . he was told about a war that took place in Limonar against the blacks, many of whom were hanged in the docks of this city [Matanzas]."[43] But the effects of the rebellion went even further. In 1842, the event was remembered once again when Captain-General Jerónimo Valdés decided to issue a new slave code for the island. One of his elite correspondents stated that the "1825 uprising of the blacks of Mr. Fouquier" should be carefully considered before giving greater freedom to the slaves.[44] During the trials of the plotters of La Escalera two years later, Pablo Gangá, the mastermind behind the 1825 movement (who had then escaped execution), was sentenced to death and hanged for his participation as a leader and sorcerer in this conspiracy discovered in 1843.[45]

Other African-led revolts followed in the 1820s, 1830s, and 1840s. Most of them had a strong Lucumí presence, but Gangá and Carabalí slaves also came to be named among the most dangerous and savage of Africans, and ethnic cooperation remained relatively common. In January 1827, fifty-seven slaves from the coffee estate Tentativa killed their mayoral and administrator and rampaged across the area, causing other casualties among the neighbors. Alarm spread throughout the region as the dispersed slaves, most of them African-born, frightened the Cuban residents of neighboring estates for over a week. The day after the uprising began, three planters of the vicinity addressed a letter to the highest local authority, Colonel Rafael O'Farrill, complaining that the slave code needed to be made effective, "especially on the coffee plantations, where thousands of Negroes and very few whites all live together within a small space of land."[46] This rebellion also presented a new problem for authorities and planters. After a large number of troops began patrolling the region, many of the rebel slaves who remained at large decided to kill themselves in acts of collective suicide. Two days after the uprising was suppressed, the owner of Tentativa, Gabriel Lombillo, complained that eighteen of his slaves had been found hanged, while the militia had shot five others.[47]

On October 22, 1827, the slaves of the coffee plantation El Carmen attempted to assassinate their mayoral, Ramón Viera.[48] According to many testimonies, some slaves decided to ambush Viera after he whipped a young Congo slave named Nepomuceno without a good reason. The slaves, led by Simón Mina, Celedonia Mandinga, and Ventura Congo, pretended that one of them had committed suicide by jumping into a well; they intended to take their revenge on the mayoral by throwing him into the well when he went to look. He must have sensed the danger, because he decided to ignore the calls for help. Instead, he ran inside the main house, where he hid until the militia sent from Güira de Melena arrived. The ringleader of the attempt, Simón Mina, declared later that he and his co-conspirators had decided to get rid of the mayoral because he "used to give them a lot of whipping, as water falls from the skies."[49]

Lucumís struck again near Havana in September 1832, cementing their reputation as troublesome slaves. Seventeen recently arrived Lucumí slaves who were bound to the sugar mill Purísima Concepción rose against their mayoral and escaped to the woods. Their revolt seems to have been inspired by a suggestion made in "their language" by their fellow slave Manuel Lucumí, who convinced them to escape from the plantation and start a new life in the mountains.[50]

Problems with Lucumí slaves seemed to grow worse every year. In 1833, a cholera morbus epidemic broke out in the western part of the island. Throughout the plantation region, authorities, slave owners, and overseers took all sorts of precautions to keep the epidemic as far as possible from their slaves.[51] Francisco Santiago Aguirre, owner of the coffee plantation Salvador, west of Havana, was no exception. He decided to give his slaves the best possible treatment in order to protect them from the disease, locating them all together in a safe space and reducing their working hours. His seeming generosity prompted one of the biggest slave rebellions ever to happen in Cuba. Three hundred and thirty slaves out of 375—most of them Lucumís—began a tumultuous revolt that caused irreversible damage to neighboring plantations and in the town of Banes.[52]

Early in the evening, the slaves began beating their drums and talking to each other in the Lucumí language. According to the mayoral of Salvador, Diego Barreiro, calls for freedom and the Lucumí song for meeting, known as the "Ho-Bé," were heard all across the estate.[53] The "Ho-Bé" was followed by a Lucumí song of war, the "Oní-Oré," which was answered by the slaves with "O-Fé." The leaders, all Lucumís, threatened anyone who refused to join them. One of the few slaves who escaped the tumult, Eusebio Congo, declared that the Lucumís were beating their drums, singing, and dancing.[54] They were also wearing colorful clothes. One of the leaders, the contramayoral, Luis Lucumí, wore a woman's dress and hat.[55]

During subsequent interrogations, the public prosecutors were forced to address the slaves by both their African names and their new Christian names. Some of the captured slaves, among them Eguyoví and Ayaí, recounted during their interrogations how their comrades had honored their military traditions during the revolt and its aftermath. They spoke of how Valé had carried the wounded Ochó on his shoulders during their retreat and how Ochó had begged Valé to kill him rather than try to save his life. They also recalled that once their situation was desperate, Valé finally decided to shoot Ochó, putting a pistol inside Ochó's mouth and ending his life with a bullet.[56]

In 1834, Lucumís rebelled on the sugar mill San Juan de Macastá, not far from the Salvador. This time, the rebellion was inspired by an excess of work. After the revolt was put down, the mayoral declared that the slave drivers had led the rest of the slaves in an attempt to kill him. He also testified that the slaves communicated in Lucumí.[57] In this case, all the slaves supported the mayoral's testimony: they declared that the mayoral wanted to make them work all day and that he was not allowing them to cultivate their own conucos. Tomás Lucumí, the contramayoral who led the rebels, declared that the mayoral had threatened them with reducing their rations because he believed that they were all very fat.[58] This threat and the impossibility of caring for their conucos seem to have been the main reasons for the uprising.[59]

The next year, four Lucumí slaves died in two intense days of fighting and negotiation at the sugar mill La Magdalena in the Santa Ana jurisdiction in Matanzas. The slaves had been recently bought and had been compelled to work after being forced to look at the bodies of two of their companions who committed suicide the day before.[60]

On June 18, 1837, after an attempt to punish one of the Lucumís under his control, Guillermo Monroy, the mayoral of the sugar mill La Sonora, realized that his action had been a mistake. The slaves, tired of Monroy's violent behavior, had been waiting for his next excess. In preparation for taking their revenge, they had been hiding sticks, rocks, knives, and machetes in the bushes. When Monroy pushed Esteban, one of the recently arrived Lucumís, Esteban resisted, calling to his comrades in Lucumí: "Companions, do not run away. . . . What can whites do against us? Let's fight them."[61] The rest of the slaves began screaming in "arrhythmic ways" and furiously attacked all the whites on the plantation. Another of the new slaves, Fermín Lucumí, stated that the conspirators had planned to rebel when, "five moons ago," the mayoral had broken a slave's head open.[62] Although the rebellious slaves did not kill anyone, the three leaders of the revolt—Esteban, Martín, and José, all Lucumís—were sentenced to death. They were executed by firing squad on the morning of November 10, 1837, in the yard of the La Punta fortress, next to the Atlantic waters they had been forced to cross only a couple of months earlier.[63] During the legal process that led to this outcome, most of the captured slaves were forced to testify through an interpreter and under an oath to their own "Lucumí God." Not surprisingly, the attorneys appointed to defend them argued that the rebels were all in a "state of barbarism" and that in their countries of origin, no one knew "the laws and considerations that men should observe in society."[64]

Less than three months later, on September 10, 1837, twenty-five Lucumí slaves confronted their mayoral and locked themselves in one of the houses of the sugar mill San Pablo, near the village of Catalina de Güines. Then, armed with machetes, they fought back against the white employees of the estate. One was killed in the battle, while two others escaped and hanged themselves in the forest in another case of multiple suicide.[65]

By this point, African-led revolts were becoming familiar events in the Cuban countryside. Newly arrived Lucumís were almost always at the center of the rebellions. They struck once again in May 1839. This time, the location was the sugar mill La Conchita, in the jurisdiction of Macuriges, not far from Matanzas. In this case, too, the slaves needed a Lucumí interpreter to offer their testimonies during the subsequent trial.[66]

Other Lucumí rebellions followed. On June 12, 1840, ten Lucumí slaves escaped from the sugar mill Banco, in Güines, and attacked a militia force commanded by Lieutenant Inocencio López Gavilán. After a fierce and bloody battle, three slaves lay dead and one was injured, while López Gavilán himself was wounded three times. The rest of the rebel slaves escaped.[67]

Cubans who lived in the vicinity of Havana had their own firsthand experience of Lucumí rebellion. On October 8, 1841, nineteen recently arrived Lucumí slaves who were engaged in work on one of the most magnificent palaces of the city stopped working and defied their mayoral and master, only to be ruthlessly repressed by the soldiers of Havana's garrisons.[68] Valentín Toledo, the mayoral, and Domingo Aldama, the slave owner and the richest man in Cuba at that time, later declared that the rioting slaves had brandished sticks and rocks while beating their buttocks and touching their genitals. On Aldama's orders, the troops had attacked, killing six slaves and wounding seven. Aldama declared that he and his men had tried to persuade the slaves to surrender. Unfortunately, neither he nor the soldiers had realized that the just-landed slaves were unable to understand Spanish and therefore could not comply with their orders. Two of the survivors, Nicolás Lucumí and Pastor Lucumí, confirmed, with the help of an interpreter, that they had not understood a single word said by the whites who had seized them.[69]

Lucumís next left their mark on the sugar mill La Arratía. Forty-two slaves—most of them Lucumís—beat their mayoral and other white employees of the plantation. Armed with rocks, sticks, and torches, they burned down one of the plantation's main buildings and briefly took control of the estate before escaping while yelling, "Kill the whites!" Five of the leaders of this revolt were captured and executed soon after by the Military Commission.[70]

Though the new slave code sponsored by Captain-General Valdés was promulgated in 1842, 1843 turned out to be a more dangerous one for the Cuban authorities and whites who lived on the western side of the island. Between March and November, some of the biggest uprisings ever seen in the New World unfolded in the plantation area around Matanzas and Cárdenas. Again, Lucumís led the first of these revolts, which broke out at the sugar mill Alcancía in March. The British consul, Joseph T. Crawford, who had informers in the area, related the events to the earl of Aberdeen a few days later, remarking on the central role of the Lucumí slaves: "On the 26th ultimo at two points close to each other, Bemba and Cimarrones in the district of Matanzas, upon five estates contiguously situated, insurrection broke out amongst the Negroes. They were all of the Lucumie nation and are famed for being the most hardy of the Africans, warlike in their own country and the most hardworking here."[71] As Crawford also wrote, the Lucumí rebels set fire to buildings on some neighboring plantations, damaged the railroad works between Júcaro and Matanzas, and killed at least five whites before being defeated by the Spanish army. The number of African casualties was exceptionally high. According to Crawford, around 450 slaves perished, either by being shot or executed or by committing suicide collectively.[72] During the uprising, the slaves beat their drums and showed their skills in warfare, bringing back memories of Guamacaro in 1825 and Guanajay in 1833. They moved "in military order, clad in their festival clothes, with colors flying, and holding leathern shields."[73]

In May, Domingo Aldama had to face two other revolts led by Lucumí slaves on his sugar mills Santa Rosa and La Majagua, both of which were situated in the jurisdiction of Sabanilla del Encomendador. In June, there were disturbances at the Ácana and Concepción plantations, and Guamacaro was the stage for the rebellion of more than three hundred slaves from the sugar mill Flor de Cuba. Unfortunately, these revolts were not very well documented, but they seem to have been dominated and led by African-born slaves. In July, Lucumís from the sugar mill La Arratía rose again, damaging their owner's property and spreading panic among the neighboring planters. This time, more than forty slaves were involved in the revolt.[74]

The last—and probably also the biggest—slave revolt in this sequence began on November 5, 1843, in the district of Sabanilla, near Matanzas. That day, slaves from the sugar mills Triunvirato and Ácana revolted, killing some whites and burning down mill houses and buildings on both plantations. After some hours of rampage and combat against the soldiers sent to repress the revolt by the governor of Matanzas, the rebels were defeated on the land of the sugar mill San Rafael. Fifty-four slaves were killed in the battle, while sixty-seven were captured. More than three hundred slaves, most of them Lucumís, had provoked one of the most dangerous uprisings ever witnessed on the island. Less than a month later, the prevailing alarm among the region's slaveholders was transformed into outright rage against their slaves. A widespread conspiracy organized by free blacks and slaves was uncovered, and its plotters were repressed, provoking some of the most bloody scenes in the history of Cuba. The repression of La Escalera had begun.[75]

The African Element

It is a fact that most of the revolts that occurred on Cuban plantations were planned and led by African-born slaves who had recently arrived on the island. Their familiarity with war is repeatedly mentioned in the records of their interrogations and in the letters and documents produced by colonial officials. The military experience of African-born slaves was no secret from Cuban authorities, planters, and the general public. Most of the Cuban population was aware that Lucumís, Gangás, Carabalís, Minas, Ararás, Mandingas, and Congos were familiar with guns, horses, and even with some Western combat strategies that they had copied from the Europeans in Africa.[76]

Slaves' African practical knowledge often proved decisive when they began plotting insurrection. According to some contemporary accounts, Gangá slaves were frequently embroiled in all sorts of conspiracies and revolts. Despite his warrior heritage, however, Martín Gangá declined to participate in the plot organized in Guamutas by Pancho Peraza Criollo in 1843 and 1844. He recalled how Peraza had urged him, "Martín! In your country your people are accustomed to make war, and consequently you should not be afraid here." Martín distanced himself from that warrior past when he answered that "he had come here when he was a very small child and that consequently he did not know about these things."[77]

In the early 1840s, following the shocking sequence of Lucumí slave revolts, Cubans generally believed that Lucumís did not want to be slaves and that they were ready to fight until they triumphed or perished.[78] Alejandro Gangá, a slave on the sugar mill Coto, near Matanzas, told prosecutors that his comrade Pedro Carabalí was always saying that "the Lucumís were very brave and that they were not afraid to die."[79] Popular opinion about the courageous character of Lucumís sometimes turned into mockery of their unsuccessful military experiences. According to Tomás Criollo, his friend Marcelino Gangá had once asked him, "If the Lucumís were so daring, how was it possible that they let others capture and sell them as slaves?" Gangá said "that the Lucumís were always thinking of rebelling, only to end up killed."[80]

But war—whether in Africa or on nineteenth-century Cuban plantations—required more than ethnic solidarity and courage. Secrecy was essential for a conspiracy to succeed. For that reason, plotters made secret pacts and swore secret oaths. Many such vows were not strictly African but rather hybrid statements that combined diverse African religious beliefs with Roman Catholicism.[81] José Gangá recounted the ritual surrounding his initiation into the 1844 conspiracy of La Escalera. Once he was sure that the plotters would provide him with weapons—he made them swear that they had firearms—José prayed the Catholic credo and, after drawing a cross on the ground with a knife, swore loyalty to God and the Virgin Mary. Finally, he and the other initiated slaves had their hair cut short along their right ears.[82] Three other slaves who made their vows that week participated in a similar ritual. They all recalled that the free black Pedro Ponte had drawn a circle on the ground and a cross inside the circle; subsequently he told them to kiss the cross and to swear loyalty upon it. Finally, they had to eat soil from inside the circle to reaffirm their commitment to the conspiracy.[83]

Sacred oaths could also be used to silence those who knew of a plot but were not absolutely committed to it. This was the case of some urban slaves in Matanzas in 1844. Domingo Lucumí described how he and his fellow slaves working in the honey warehouses of Matanzas were forced by a number of free blacks and mulattos to become part of the conspiracy of 1844. He recounted that Santiago, one of the freemen, had made them take their hats off and promptly had brought a bottle of holy water. Then he had mixed some red powders with the water and had forced all the slaves to drink it and to say, "If I reveal the plot, I will fall apart." Many of the slaves initially had refused to drink the mysterious beverage, but Santiago had threatened them by putting his knife to their throats. After the ritual was over, Domingo had rushed to rinse his mouth with gin in a tavern next to the Yumurí River, in a clear attempt at physical and political detoxification.[84]

Leadership also played a significant part in the unfolding of a revolt. As we have seen, slaves in the position of contramayoral often used their post to launch some of the most daunting African-led rebellions in the Cuban countryside. The revolt of 1833 in Guanajay (organized by three slave drivers), the rebellions at the sugar mills San Juan de Macastá in 1834 and Guaycana-mar in 1840, and above all the great Lucumí uprising in the jurisdiction of Bemba in March 1843 are among the most notorious examples of the role played in some slave movements by slave drivers.[85]

Some slaves were leaders in Africa or acquired a reputation as brave men in the course of their military careers there. Frequently, slaves who had participated in Cuban revolts testified that their only motive for rebelling had been their commitment to their captains or kings from Africa. As we have seen, this was why Rafael Gangá offered his life in the revolt in which all his comrades and his captain perished in 1827.[86] On some occasions, slaves' leaders were also their relatives, as was the case with the Carabalís from the coffee plantation La Hermita, who rebelled in 1825 in Guamacaro. It was also the case with the Congos from Cayajabos, who murdered their master, Pedro Rodríguez, in November 1812.[87]

Almost every African-led movement had leaders who sometimes asserted their leadership by using titles or wearing specific costumes. On several occasions, slaves under interrogation referred to their leaders as both "captains" and "kings." A rather unusual case was that of José Antonio Ramos of Gibacoa, who was planning to become the "principal and bigger lord of his region." According to Francisco Congo, one of Ramos's subordinates, "there was not any agreement about the title [Ramos] would bear [once they won], but it was clear that he would be the recipient of all the tributes."[88]

Costumes were symbols of power and leadership. Sometimes they were stolen from their owners after an uprising had begun. During the rebellion of 1825 in Guamacaro, Cayetano Gangá stole and wore a green jacket "like those used by the lieutenants of the Spanish regular army."[89] In 1833, con-tramayoral Luis Lucumí stole some women's clothes, including a hat, from his master's house and wore them until he was shot in the chest later that day.[90] On other occasions, costumes were carefully prepared in advance and secretly kept for months. In 1844, Roque Gangá declared that he was chosen to be the king of his fellow slaves at the sugar mill La Mercedita. Another slave on the plantation, Dolores Carabalí, had been looking after his royal costume, which consisted of striped blue trousers, a white shirt, and a blue woolen cap decorated with two black stripes and three red tassels, one in the front and the other two next to the ears.[91] His companion, Gertrudis Carabalí, had her own clothing for the day she would be proclaimed queen. Her costume was composed of a "white muslin dress with short sleeves and a decorated silk bonnet with some holes, probably made by the mice."[92]

Drumming and singing were part of most of the African-led slave revolts that occurred in the early nineteenth century. Though the lyrics of most of the songs were not recorded and it is virtually impossible fully to understand the actual importance of the drums, it is a fact that singing and drumming stemmed from the slaves' African warfare traditions. The "Ho-Bé" and the "Oní Oré Ofé" songs for assembly and warfare—sung in the summer of 1833 by the Lucumís on the coffee plantation El Salvador and which were, unusually, documented by the colonial authorities—offer perhaps the best example of the significance of singing for African rebel slaves.

Another common feature of African-led revolts was the use of African languages by the rebels to communicate. After all, there was no reason why they should conspire and revolt using their masters' language, which most of them did not speak or even comprehend. The fact that Lucumí was spoken by Carabalís and also at least partially understood by Ararás and Minas made it a sort of lingua franca among West African slaves in nineteenth-century Cuba.[93]

Other African elements also characterized this sequence of slave rebellions. In many of the revolts, recruitment was based on ethnic empathy. Slaves from the same geographical regions or from similar ethnic backgrounds proved to be more likely to plot and rebel together. Their common languages, knowledge, and cultural backgrounds played a significant role in their decision to participate or not in a potentially hazardous undertaking. But recruitment was often also effected on the basis of fear. Several captured slaves testified that they were forced to follow their rebel companions against their will. Those who refused to join the rebels ran the risk of being considered traitors and thereby placed their lives in jeopardy.

Once revolts began, slaves frequently raided neighboring plantations in order to add new forces to the original rebel group. They marched, destroying and plundering, sometimes flying colored flags, often following the rhythm of their drums and carrying all sorts of weapons, from leather shields, rocks, spears, and bows and arrows to knives, machetes, and guns.

A final distinctive element that characterized African-led revolts in Cuba was the use of various sorts of magic charms by the slaves to protect themselves in combat and to guarantee their victory. Numerous African-born slaves testified to the relationship between magic powers and warfare. During the conspiracy of 1844, Campuzano Mandinga, who was considered the Great Sorcerer of Matanzas, not only became a celebrity among his comrades but also made a small fortune selling all sorts of magic devices as protective amulets for the upcoming war.[94] He, along with many other African magicians, prepared and sold hundreds of amulets before the conspiracy was exposed.

As cultural leaders, sorcerers often became the ringleaders of conspiracies and served as military captains in African-led revolts.[95] African sorcerers were respected and feared. Campuzano was a kind of spiritual leader for many slaves of Matanzas, and Pánfilo Congo, a slave from the sugar mill Atrevido, was feared by all those who knew him, as his menacing nickname "Dream Swatter" suggests.[96]

Most magic amulets were designed either to stop bullets and sabers or to make their bearers invisible to the eyes of their enemies. While being interrogated, Antonio Criollo described the amulet of his friend Agustín Mina, who had "wrapped in a handkerchief frogs, snakes, and many other things."[97] Another Creole slave, a coach driver for a neighboring sugar mill owned by Tomás Adan, offered a similar description. He recalled that Agustín and another slave called Basilio had assured him that their amulets protected them against bullets and machetes. Agustín, the slave said, had carried a bag filled with "many bones, snakes' heads, frogs, chameleons, and powders of almagre."[98]

Authorities, too, believed in the strength of African magic charms and were keen to eliminate them, but they were not sure how to do so. Throughout the first half of the nineteenth century, slave owners and authorities repeatedly complained about African beliefs in what they called "witchcraft." Although slave owners rarely made religious education a high priority, they considered the Christian indoctrination of the slaves to be a very important matter. As late as 1848, the bishop of Havana, the captain-general of the island, and the minister of finance were embroiled in a four-year debate about how to change African slaves' beliefs.[99]

In 1844, the Military Commission prosecutor, Apolinar de la Gala, gave his opinion about the issue of African witchcraft. He began by calling for all relevant authorities to take action on this important issue. He complained about the large number of "this kind of soothsayers" who were roaming the island and about the even larger number of "believers" in their charms, who "all came from the same part of the globe."[100] Gala lamented the fact that even once the Cuban government and the island's planters decided to act to eliminate the slaves' beliefs in magic, "it would take several years to produce the healthy and desired effects [needed] . . . for these people to renounce their native beliefs." Finally, aware that slaves saw amulets as giving them a psychological advantage, Gala called for amulets to be obliterated, describing them as a "diabolical inspiration" and "a formidable weapon."[101]

Conclusion

The overwhelming majority of slave revolts that occurred in Cuba during the Age of Revolution were not induced or provoked by external factors. Rather, they grew out of personal experiences—cultural, military, religious, and so forth—rather than revolutionary ideas imported to Cuba from elsewhere. In the context of the master-slave relationship, homicides can be also considered as directly stemming from African traditions. In West and West-Central African warfare, many African-born slaves considered killing their enemies to be a completely natural and accepted way to proceed.[102]

There were, indeed, movements organized and led by free blacks and mulattos, mainly Creoles, that aimed to destroy the colonial system of slavery. These movements were mostly urban-based and included large numbers of free people among their leaders. La Escalera, for example, has been considered a slave conspiracy since it was uncovered late in 1843. A quick examination of the list of La Escalera conspirators reveals, however, that slaves made up only 25.45 percent of the convicted plotters, while free blacks and mulattos accounted for 71.09 percent.[103] A conspiracy in which only a quarter of the participants were slaves can hardly be labeled a "slave conspiracy."[104] Rather, like the conspiracies of Guara in 1806 and Aponte in 1812, La Escalera was a well-planned plot organized by urban free blacks and mulattos—mostly Creoles—who were aware of the sympathy that African slaves would feel for their cause and who did not hesitate to invite them to join their movement. Its leaders were also aware of the fact that the large number of African slaves living on the island could incline the balance of force toward their side once the war had begun. Slave support was too significant to be wasted.

Not surprisingly, the slaves involved in both plots were in the vast majority Creoles or ladinos who had lived long enough on the island to be able to speak Spanish and, in many cases, to exchange their African surnames for the surnames of their masters.[105] In La Escalera, no one African group was preponderant among the rebels. The lengthy list of plotters was balanced between Lucumís, Congos, Carabalís, and others. La Escalera's slaves were predominantly urban and Creole, although slaves from the countryside— very often coach drivers or slave drivers—also seem to have been willing participants in the plot.[106]

The slaves involved in the conspiracy of Aponte were also mostly urban. They were cabildo members—especially in the cabildos Mina Guagni, Congo Masinga, and Carabalí Osso—and naturally they met inside their cabildo houses, although taverns and holiday parties were also appropriated to discuss their plans. With the sole exception of the Carabalí-led rebellion that occurred in Puerto Príncipe in January 1812, all of the Aponte-related movements were organized and led by free Creoles. The uprising at the sugar mill Peñas Altas, for example, was organized by José Antonio Aponte's companions—Estanislao Aguilar, Juan Bautista Lisundia, and Juan Barbier—and the slave drivers of the plantations. The slaves, most of them Congos, did not play any role in the leadership of the rebellion.[107]

To state, then, that most of the slave movements that took place in Cuba— and also in other places that received African slaves until well into the nineteenth century—during the Age of Revolution were the consequence of foreign influences and ideas is inaccurate and misleading. It suggests that the reasons African-born men and women had to fight their slave condition were not important or worthy because those reasons were not truly revolutionary or politically motivated. Simply put, African cultural, religious, political, and military continuities were the central influence on the slave conspiracies and revolts that took place in the first half of the nineteenth century in Cuba. Guara, Aponte, La Escalera, and some other movements that counted slaves among their participants should not be considered slave plots and revolts, since they were movements conceived by free men; they need to be studied as such by future scholars. The free colored men who tried to bring about social change for themselves and for their enslaved friends and companions, many of whom lost their lives in the process, deserve all the credit for their efforts. That they fought against slavery reveals their dedication and bravery; that slaves trusted and followed them indicates that they did not lack leadership skills and popular appeal among the masses of slaves.

More work is needed on the many virtually unknown Cuban slave revolts of the early nineteenth century to correct erroneous assumptions about the character and aims of the thousands of African slaves who very likely lost their lives fighting slavery without having ever heard a word about the French or the Haitian revolutions or about the British abolitionist movement. Future studies will also help to reveal the magnitude of the sacrifices made by many free colored men and women who ended their lives on the gallows or in the darkness of dungeons for daring to challenge the Spanish slave system.

3

Marronage

He broke his iron chains into pieces in the woods, smashing them persistently with a rock.
—Deposition of Alejandro Congo, Candelaria, August 21, 1844

On April 21, 1844, public prosecutor José del Mazo wrote to the governor of Matanzas, Antonio García Oña, about the escape from custody of the most important slave leader in the legal proceedings Mazo was conducting not far from the provincial capital. Mazo expressed both surprise and anxiety at the escape, writing, "The most important prisoner of this process, the Negro Ramón Gangá, ran away from the sugar mill owned by Don Domingo Riverol. He escaped despite having shackles on one of his feet, chains on both, and after bringing down a wall of bricks. . . . I investigated the event afterward, and it seems that he did not have any sort of help to flee, a fact that has really mortified me."[1]

Marronage was one of the most feared forms of slave resistance in the New World. Only the possibility of a slave revolt could match the panic that runaway slaves provoked. Throughout the Americas, maroons managed to perturb colonial authorities; on more than one occasion, they forced local authorities to comply with their requirements. And, with the exception of the Haitian Revolution, marronage was the only successful overt form of slave resistance in the Americas.

Since the mid-seventeenth century, runaway slaves had established communities in isolated locations, creating alternatives to life on the plantations.
In Brazil, for example, runaway slaves formed large communities in the
northeastern provinces of Pernambuco, Sergipe, Alagoas, and Bahia, repeatedly defeating Portuguese colonial armies sent from Recife, Salvador, and
Rio de Janeiro to destroy them. In Dutch Guyana, they created such a strong
network that at one point they seemed about to succeed in expelling the
Dutch from the territory known today as Surinam. In Jamaica, led by their
leaders Cudjoe and Quao, the Leeward and Windward maroons even managed to sign two peace treaties with the British colonial authorities between
1738 and 1739 to stop the war that had plagued the colony since the British
took over the island from the Spanish in the mid-seventeenth century.[2] Many
pages have been written about runaway slaves in the Americas, their reasons
for escaping, and their techniques for survival. Some of the works that portray their bravery and suffering have become best sellers, such as Esteban
Montejo's Cimarrón and Alex Haley's Roots.[3] Other, more objective books
and articles have explored in depth the most important questions related
to the lives and fortunes of runaway slaves in the New World. These studies
have assessed many aspects of agrarian production in runaway communities, from their tools to their diverse crops.[4] Scholars have also called attention to the significance of maroons' knowledge of local geography, stressing
the advantages of knowing the woods and mountains for everything from
increasing the probability of escaping slave hunters[5] to the exploitation of
natural resources, such as minerals or fluvial ways.[6]

Other scholars have uncovered the different ways in which runaway slaves
re-entered the "civilized" society from which they had escaped. Gloria García
has shown that maroons constantly went back to the plains for any number
of reasons. Maroons often preyed upon their former plantations, as Matthias
Röhrig Assunção and Wim Hoogbergen have shown, and they also traded
with their comrades who remained enslaved.[7] In some cases, as in the quilombo (runaway slave village) of the Oitizeiro, south of Salvador da Bahia
in Brazil, the relationship between maroons and slaves reached the point
where maroons were being hired by coiteiro slaves to work their gardens.[8]
The complex relations of production in the Oitizeiro defy the traditional
scholarly assumption that quilombos—or palenques, as they were called in
Cuba—remained totally isolated from the rest of Cuban society. In some

cases, maroons not only traded with and visited plantations but also organized celebrations and conspiracies alongside urban slaves, as was the case in 1814 and 1826 in Bahia and between 1843 and 1844 in the hinterland of Matanzas, Cuba.[9]

The internal forms of organization of runaway slave communities in the New World and their tactics of defense and attack have been studied in the last few decades. What planters called the "maroon issue" and its categorization have received special attention, as marronage was one of the most distinguishable and regular forms of slave resistance in the New World. Scholarly interest has led to the publication of books and articles assessing marronage in nearly every corner of the continent and even in Africa. The 1973 collection Maroon Societies, edited by Richard Price, is a a seminal international collaboration that explores the history of runaway slaves and their lives.[10] Maroon communities in places as diverse as Brazil, Dutch Guyana, Jamaica, and Cuba are analyzed as examples of how runaway slaves organized their lives in the woods. This study and others stress how slaves had a vast and highly developed geographical knowledge of the areas in which they lived. Their guerrilla tactics and their ways of obtaining provisions, as well as their relationships with slaves on nearby plantations, have all been key topics of discussion in recent decades.[11]

Cuban scholars have also produced several works on this theme. Since the studies of Fernando Ortiz and José Luciano Franco, however, most works on marronage have echoed official discourse about the bravery and nobility of the maroons and have lacked depth and serious research.[12] Strong political interests motivate this approach. Comparisons between maroons and Cuba's post-1959 revolutionary soldiers, as well as between palenques and the rebel anti-imperialist island, signal the condition of maroon studies in Cuba. According to historians such as José Luciano Franco, Ramiro Guerra, and Pedro Deschamps Chapeaux, maroons constituted the most serious threat Spanish colonial authorities ever faced prior to the Wars of Independence. These scholars have glorified the maroons' "permanent" rebelliousness, but their positions have sometimes proved to be poorly argued. In trying to exalt the image of runaway slaves, for example, Guerra wrote—without further developing the idea—that "the Negro, as a human creature, always wanted to be free even when the price of this freedom could be his own death, in the palenque or anywhere else."[13] In his analysis of marronage and slave imports to Cuba, Pedro Deschamps Chapeaux went even further, maintaining that he did not understand why, despite the "permanence of marronage" in Cuba in the first half of the nineteenth century, "the imports of slaves kept their increasing rhythm, according to the exigencies of the plantation system." In his attempt to give greater significance to the maroons' place in the Cuban colonial system, Deschamps Chapeaux even suggested, quite naively, that authorities, merchants, and planters were at times so frightened of maroons that they considered abandoning African slave labor altogether.[14]

Gabino La Rosa Corzo's academically rigorous work is an exception to the general rule. Relying mostly on archival sources, La Rosa Corzo has produced two serious and outstanding studies on Cuban marronage. His second book, which treats the maroons of the east side of the island and is enriched by the accurate addition of archaeological evidence, is perhaps the best work on the subject of Cuban maroons to date. Nevertheless, he has also reproduced the idea of the brave, invincible maroon, as well as the even more debatable myth that maroons represented an actual danger to the colonial status quo in Cuba.[15]

Authorities

In 1796, the Royal Council of Agriculture and Commerce of Havana approved and issued the *Nuevo reglamento . . . de cimarrones* (New Regulations . . . for Maroons), a code of rules for dealing with the swelling maroon population.[16] Despite the existence of previously established rules for dealing with maroons, before 1796 they were hunted down by freelance or locally employed rancheadores, who either disobeyed or ignored the existing royal dispositions. In 1796, for instance, the countess of San Felipe y Santiago, the local authority of the village and county of Bejucal, had a large number of freelance slave hunters under her protection.[17] Marronage was controlled to a certain degree in Bejucal due to the limited number of slaves and the efficiency of these men and their well-trained hounds. Yet blacks—Africans and their descendants—were rapidly overtaking whites as the largest racial group in Cuba.[18] As a consequence, marronage began to be a relatively serious problem in the plantation area around Havana and Matanzas and in the mountainous eastern region of the island. The experience of the Haitian Revolution and the success of the Jamaican maroons in negotiating treaties with the British influenced the authorities' belief that the maroon question needed to be tackled as soon as possible.

The code divided runaway slaves into two main groups: apalencados and simples.[19] The term "simple" was used for slaves who were captured within a certain distance of their home state who did not possess a paper that authorized them to travel. "Apalencado" denoted runaway slaves who formed groups of seven or more. The code also explained the responsibilities of planters and rancheadores. The former were obliged to submit monthly lists to local authorities of their fugitive slaves, and the latter's rewards were fixed in accordance with the difficulty of each slave's capture.[20] The code prohibited rancheadores from mistreating runaway slaves once they were captured while conceding that it was not possible to apply the same rule during the "moment of attack," thus leaving the slaves' lives in the hands of their hunters.[21] The code specified where captured slaves should be kept until their masters claimed them, stated that unclaimed slaves would be employed on the public works of cities, and enumerated penalties for those who did not rigorously observe its rules.

The *Nuevo reglamento* of 1796 was periodically modified during the first half of the nineteenth century. Adjustments were made according to the requirements of the historical moment and were mostly prompted by the steady increase in imports of African slaves. Usually, modifications addressed small administrative problems, such as slave hunters' salaries and the destination of captured slaves. Yet the efforts of the Cuban authorities were not restricted to promulgating new versions of the code of 1796. Marronage and its associated problems created an ongoing dilemma. The authorities' frequent correspondence with the jurisdictional captains in charge of slave hunters across the island reflects the importance they gave to the maroon problem, particularly in social or economic terms.

When Francisco de Arango y Parreño drew up the *Nuevo reglamento* on behalf of the Cuban authorities in 1796, he was well aware that it was in consequence of his having written another document four years earlier, the "Discurso sobre la agricultura en Cuba y medios para fomentarla" ("Discourse on Agriculture in Cuba and the Means to Promote It").[22] For Arango and his fellow planters and merchants, the opportunity of replacing the neighboring French Saint Domingue was too tempting to ignore. Once the collapse of Saint Domingue was a fact, Cuban planters and merchants put pressure on the king of Spain to obtain permission to import slaves directly from the African continent. Several petitions based on Arango's "Discurso" were made, and the results were almost immediate. Charles IV agreed to introduce a massive number of Africans to Cuba to work on the sugar and coffee plantations that were spreading west and east of the cities of Havana and Matanzas. Soon thereafter, maroons, who had not previously been a problem, began to menace the harmony of the successful Spanish colony. Fearing that the number of maroons in Cuba would multiply, Captain-General Luis de las Casas reduced the number of rancheadores who traveled to Jamaica in 1796 to hunt down maroons there. The success and growth of maroon communities in the Jamaican mountains and the outbreak of civil war in Saint Domingue were too close in time and space to be disregarded.

As a result, the maroon issue was addressed before it could become a real threat to the Cuban elite and its plantation-based economy. The code of 1796 helped to institutionalize the practice of hunting down maroon slaves, and it established a body of rules that all citizens had to follow. Authorities from Havana and Santiago de Cuba consequently developed a well-designed hierarchical system of communication to control the different aspects of slave hunting across the island. The royal council in Havana was at the top of the ladder, with jurisdictional captains, plantation owners, and slave hunters all compelled to report monthly about the maroon situation in the areas under their command and supervision. The governors of Santiago de Cuba, Puerto Príncipe, Las Cuatro Villas, and Matanzas were required to report any news about maroons directly to the captain-general.

Although colonial authorities mainly dealt with matters beyond the functions of the captains of jurisdiction, they sometimes became involved with the maroon problem in very practical ways. On August 18, 1819, for example, a few selected members of the government of Havana and the Cuban Royal Council of Agriculture and Commerce held a meeting at the captain-general's palatial headquarters. They gathered that summer morning to consider the matter of the palenques formed by runaway slaves in the vicinity of the "first village" of the island, Baracoa.

Palenques and their inhabitants were not an uncommon discussion topic for colonial authorities. Quite often, communications and letters sent by slave hunters, slave owners, and local authorities were read and discussed both in the ayuntamiento (city council) and in the royal council. That day, however, a particularly significant event brought them together in the governor's palace. Three months earlier, a group of thirty-one apalencados armed with guns and machetes had come down from the mountains and had gathered in the outskirts of the village of Baracoa on the eastern end of the island. These maroons appeared in a sparsely populated area and told the local residents that they were bearers of an official order issued by the governor of Santiago de Cuba, José Eusebio Escudero, which obliged them to open a road through the mountains to that city. It is difficult to know why, but it seems that most of the villagers believed what the group of runaway slaves told them. Taking advantage of the success of their ingenious lie, the maroons began crossing the region freely, arousing alarm among nearby residents. After seeding terror and stealing fourteen slaves from local plantations, they retreated to the mountains.

The leader of this group was a slave named Vicente Pérez, known as Cobas. Apparently, he and the governor of Santiago de Cuba, José Eusebio Escudero, had previously made a verbal agreement that the maroons would stop their raids and disband the palenque in return for a royal pardon. But rather than heading to Santiago de Cuba to surrender to the governor, Cobas and his men dramatically defied that understanding by their actions in the mountains. According to letters received by the secretary of the ayuntamiento of Havana, Cobas's failure to comply with the agreement changed "entirely the aspect of the [marronage] matter."[23]

Escudero sent a printed communication to Havana, explaining the circumstances surrounding his agreement with Cobas and mentioning in passing another attack that had happened under his jurisdiction four years earlier: "All together and with the speed of lightning they cross through the forests without problems, twenty, thirty, or forty miles, to fall like hawks on the neglected landholders from whom they steal everything. . . . In such circumstances, those of their kind who run to defend their masters and mayorales are murdered or in the better cases beaten, as happened in February of 1815 when sixty or seventy blacks came out of their cave in the San Andrés Mountain and simultaneously, at night, assaulted and plundered three coffee plantations, killing one of their mayorales who was not able to escape in time."[24]

In April 1822, another group of maroons came down from the mountains in the western locality of Cayajabos. They sacked and plundered the area to such an extent that the captain-general had to send the armed militia to pacify the region. A few days later, when the situation was under control and some of the maroons had been captured, he recalled his men from the area.

Nonetheless, he ordered the captain of the jurisdiction to create a company of men under the command of "someone of renowned bravery, probity, and knowledge of the mountains" to pursue the "fugitive and malicious blacks that inhabit those mountains."[25]

Perhaps the most subtle question the authorities had to deal with was whether or not the maroons represented a real threat to Cuba's prosperity and tranquility. The agreement between Governor Escudero and the maroon Cobas could lead one to overestimate the number and importance of maroons in the eastern part of Cuba, but other evidence strongly suggests that marronage was not a particularly grave problem and that it was usually under a certain degree of control. Many letters and reports indicate that colonial authorities in Havana were suspicious about the actual magnitude of the maroon problem. The high cost of paying slave hunters was the cornerstone of their discussions and the source of their skepticism. Around 1816, Governor Escudero formed a troop of four hundred well-paid slave hunters to clear out the palenques of Baracoa. The attack was brutal and ultimately successful; leaders of the troop were recommended for promotion after the palenques were utterly destroyed and a group of only fifty slave hunters chased the dispersed fugitives.[26]

Despite the success of the expedition, the maroons appeared again in 1818, and one year later their actions were being carried on so publicly that Governor Escudero was forced to offer Cobas a peace agreement. The members of the ayuntamiento and the royal council of Havana were puzzled as to how a palenque populated by roughly seventy maroons could be reborn in less than two years, especially since the region's slave owners and authorities had not complained of runaways. They wondered what was happening to their money, which was evidently insufficient to suppress the maroons. The whole matter was suspicious. The ayuntamiento and the royal council were very apprehensive about the seemingly endless requests for funds received from Santiago de Cuba and Baracoa to pay for slave hunters and the maintenance of captured maroons. Of even greater concern were the actions of Governor Escudero. Whether the maroons were a serious threat or not—and the authorities in Havana were almost convinced they were not—the evils of the palenques were not as harmful as the plan for their reduction, if only because Escudero's pact with the maroon leader seemed to suggest that the authorities intended "to capitulate to the Negroes."[27]

The reservations of Havana's ayuntamiento and royal council were backed by other facts. In the reports and journals of the most important slave hunters, months passed in which not a single palenque—and sometimes not even a single fugitive slave—was found. Rancheadores' letters and reports suggest that there were no important palenques in the mountains and thereby put to rest authorities' concerns about the maroon problem. In June 1822, Gaspar Antonio Rodríguez literally begged Captain-General Nicolás Mahy to order him to retreat, since "very little can be done when there is no group [of maroons] to be addressed."[28] Six years later, in 1828, slave hunter José Pérez Sánchez wrote in his diary, "It has been a long time since I have heard someone talking about the existence of palenques around here, and even [longer since] the maroons have damaged any property."[29]

From the point of view of the authorities and members of the most powerful institutions of the island—including local governors, the captain-general, the Havana city council, and the Royal Council of Agriculture and Commerce—maroon slaves were largely a secondary problem. The authorities' main concern was the amount of money that was spent to constrain their actions and capture them. For slave hunters, however, maroons were a part of daily life—a potential source of income, even as their dangerous captures threatened both injury and death.

Rancheadores

Rancheadores had one of the most dangerous jobs in colonial Cuba. Their skills were highly appreciated, and they were known for their effectiveness as early as the eighteenth century not only in Cuba but also across the Caribbean. In 1796, Jamaican authorities sent a mission to Cuba with the aim of hiring some rancheadores and their dogs to track down Jamaican maroons.[30] This mission was a success: shortly after the rancheadores' arrival on the island, maroons began to surrender, reestablishing the peace.[31] But even prior to 1796, slave hunters' skills were widely recognized and they were being exported to neighboring territories. There are references to an expedition to the Spanish Mosquito Coast to "appease" the native Indians. Contemporary historian Robert Charles Dallas commented on the desperate situation, when the Spanish authorities decided to bring in "thirty-six dogs and twelve chasseurs" to hunt down the indigenous inhabitants.[32] He continued, "These auxiliaries were more formidable than the finest regiment of the most warlike nation could have been; and from the time of their being employed, neither surprise nor ambush annoyed the troops; the Spaniards soon succeeded in expelling the Musquito Indians from the territory on the coast, and quietly occupied Black River, Bluefields, and Cape Gracios a Deos."[33]

Life for the rancheadores changed a great deal after the promulgation of the Black Code of 1796. Hitherto, they had worked as freelance bounty hunters or had been employed by local authorities or slave owners. After the promulgation of the 1796 code, they were required to obey central orders issued from Havana, to send monthly reports of their actions, and to keep up a regular correspondence with the royal council in Havana or with the closest lieutenant-governor. Thanks to these requirements, it is possible today to read rancheadores' daily accounts of their efforts, learning how they worked and how many men were usually under their command. As slave hunters, they had to follow the paths of runaway slaves; thus their lives in the mountains and swamps did not differ greatly from the lives of those whom they chased. They had to deal with unpredictable weather, know how to survive with a minimum of resources, and be prepared to face death from disease or at the hand of a maroon.

There are many descriptions of Cuban rancheadores, all given by whites who had appreciation for the work they did. In 1803, Dallas reported, "The chasseur has no other weapon than a long strait muschet, or couteau, longer than a dragoon's sword, and twice as thick, something like a flat iron bar sharpened at the lower end, of which about eighteen inches are as sharp as a razor. . . . These men, as we have seen, are under an officer of high rank, the Alcalde Provinciale, and receive a good pay from the Government, besides private rewards for particular and extraordinary services. They are a very hardy, brave, and desperate set of people, scrupulously honest, and remarkably faithful."[34] The motivations of rancheadores were much more complex, however, than this romantic image suggests. Most of them were relatively poor men who hunted down runaway slaves in order to feed themselves and their families, and their responsibilities were frequently demanding and unpleasant.[35]

The most notorious slave-hunter captains of the period, Matías and José Pérez Sánchez, Gaspar Antonio Rodríguez, and Francisco Estévez, all reported on the difficulty of their lives in the mountains, and all had to suspend their activities at some point due to illness among their men or their own health problems. Some of these captains were forced to look after their men almost continuously. In Gaspar Antonio Rodríguez's group, men were frequently nursing wounds or recovering from illnesses. Rodríguez used to visit these men every month, and when it was impossible for him to do so, he would send his second-in-command to check on the state of their health and to provide anything they needed.[36]

Sometimes, encounters with maroons resulted in slave hunters being wounded or killed. More than once while crossing the mountain range of Las Guacamayas, west of Havana, rancheadores remembered a freelance colleague who had been killed some years before in a place known as the Cueva del Indio. In March 1815, while checking out the remains of a palenque in the cave, Matías Pérez Sánchez briefly mentioned the "palenque where they killed Juan de Acosta."[37] Five years later, in July 1820, Rodríguez, standing in the same place, mentioned it as the "spot where the maroons murdered a young man who was chasing them."[38]

Just three months after writing this comment in his journal, Rodríguez had a firsthand experience of such violence. On October 4, 1820, he and his men attacked a group of maroons in a place called Sabana de Maíz. Most of the maroons escaped immediately, leaving behind their weapons and their provisions. One of them—the only one carrying a pistol—shot at the slave hunters before fleeing, and the bullet struck the leg of Angel Ribatón, one of Rodríguez's men. What followed was truly an odyssey for the slave hunters. They first carried the wounded man on their shoulders to the village of Candelaria, a two-day walk through the mountains. The only doctor they found was afraid to amputate Ribatón's leg, although he believed that doing so was the only way to save the man's life. On October 8, a council of medical doctors agreed to perform the operation. In the meantime, Rodríguez and his men were dashing to the larger villages of the region in search of medicine for Ribatón, who, far from getting better, was looking worse every day. On October 24, Rodríguez was on the trail of eight maroons when he received notice from Candelaria of Ribatón's declining health and the doctors' differing opinions about what to do. He abandoned the maroons and, after rushing to the nearby villages of Cayajabos and San Marcos to buy antiseptic thread and medicines and to get two other doctors, he headed to Candelaria in a last attempt to save the life of his man. On November 17, Ribatón died.[39]

Most captains of rancheadores met with similar situations in the course of their careers. Fernando de Osuna had one man under his command who was badly injured by a bullet that hit his shoulder blade and chin.[40] One of José Pérez Sánchez's men suffered a similar injury in 1828, except that his wounds were caused by two big rocks rather than by bullets.[41] Accidents were frequent and sometimes grave. Another of Pérez Sánchez's men lost part of one of his hands when his rifle exploded after he pulled the trigger.[42]

The job of rancheador required certain skills, the most important of which was a sound knowledge of the mountains. Lifetime residents of the mountainous regions, some rancheadores began their careers by serving as guides for other slave hunters. José Pérez Sánchez was one of them. Before taking charge of his own troops, he served under Rodríguez in the late 1810s and early 1820s.[43] The required level of familiarity with the mountains was not restricted to knowledge of remote paths or fast shortcuts. It was also necessary to know how to survive during the long rainy season while short on food and plagued by insects and other dangers. Dallas wrote of the harsh conditions the rancheadores faced: "Such is their temperance, that with a few ounces of salt for each, they can support themselves for whole months on the vegetable and farinacious food afforded by the woods. They drink nothing stronger than water, with which, when at a distance from springs, they are copiously supplied by the wild pine, by the black grape withes, which are about two inches in diameter, and the roots of the cotton-tree."[44] Though he recognized the difficulties of survival in the mountains, Dallas's opinion of the rancheadores' hardships lacked understanding of their lives beyond the plantations of Bejucal. He wrote that the "greatest privation" suffered by the rancheadores while on duty was that they were forbidden to smoke cigars.[45]

In addition to a keen familiarity with the geography of the mountains and a knowledge of how to survive there, rancheadores needed dogs to accomplish their gloomy task. Raising, training, and using dogs occupied a significant portion of their lives. Dallas wrote of the relationship between rancheadores and their dogs: "These people live with their dogs, from which they are inseparable. At home the dogs are kept chained, and when walking with their masters, are never unmuzzled, or let out of ropes, but for attack."[46] He also noted that rancheadores frequently beat their dogs unmercifully, using the "flat sides of their heavy muschets."[47]

Rancheadores used three breeds of dogs for hunting. The first were very big, with "obtuse noses" that were "rather squarer set," which helped them to hunt down runaway slaves. Presumably crossed with mastiffs, these dogs were very tall and consequently imposing, although they tended to be less fierce than the standard dogs used for hunting. The second breed of dogs used to track down runaway slaves were known as "finders." Dallas mentioned that their sense of smell was very keen and that they were always "sure of hitting off a track." In the only surviving engraving of a Cuban rancheador, the man stands with his two main dogs and a finder. Finders were also used to hunt small animals, such as wild hogs, to be eaten by their masters and the other dogs. "The pursuit of the game is entirely the province of the finder," Dallas wrote, while "the larger dogs, from their training, would pass a hog without notice."[48]

The most important type of dog was the third breed, the hound used to chase and capture the maroons. These animals were bigger than finders but smaller than the mastiff-crossed dogs. "This breed of dogs, indeed, is not so prolific as the common kinds, though infinitely stronger and hardier," Dallas wrote, continuing, "The animal is the size of a very large hound, with ears erect, which are usually cropped at the points; the nose more pointed, but widening very much toward the after-part of the jaw. His coat, or skin, is much harder than that of most dogs, and so must be the whole structure of the body, as the severe beatings he undergoes in training would kill any other species of dog."[49]

Since making a profit in their business required maroon slaves to be captured alive, rancheadores took care to train their dogs not to kill. The countess of Merlin remarked that rancheadores' dogs were unique in their strength, intelligence, and aversion to maroons, noting that they stopped their victims without killing them: "With an admirable accuracy and lightness, they jump on their opponents, trying to bite their ears, and once they reach this target, sink their teeth with such a strength that the pain makes the Negro succumb and give up to the *rancheador* . . . but if the Negro does not offer resistance, as happens frequently because of the fear they feel when the dogs attack them, the latter does not harm him, and instead he obliges him to walk before him, ready to bring him down at the slightest attempt to escape."[50] And, indeed, maroons feared these dogs. It is very likely that some of the dogs were not as well-trained as they were depicted by Dallas and Merlin, for they often killed their victims whether the maroons offered resistance or not.

The reputed viciousness of the hounds is often mentioned in the chronicles of the rancheadores. In January 1828, for example, José Pérez Sánchez

abstained from chasing four fugitive slaves after the manager of the plantation asked him not to, fearing that Pérez Sánchez's dogs would kill them.[51]
On November 18 of the same year, Pérez Sánchez found three slaves at a
crossroads. One of them tried to escape, so he released two of his best dogs,
which promptly killed the man.[52] More than two years later, Pérez Sánchez
released his dogs again to chase a maroon slave. Although they captured and
fatally injured the slave, four of the dogs also lost their lives in the fight.[53]
And in April 1844, the captain of the district of Camarioca wrote to the
governor of Matanzas, reporting that he had lost one of his dogs in a fight
against a maroon called Nepomuceno, who was running away after stealing
a lamb. Without expressing any regret, he stated that his other dogs killed
the slave.[54]

Despite the dogs' fierceness, maroons often managed to escape from
them or to destroy them before fleeing their hunters. During an attack by
Pérez Sánchez's men on a group of fugitive slaves in April 1828, the slaves
managed to kill all of the dogs.[55] In 1832, the dogs of Joaquín Figueredo and
his men suffered the same fate after Figueredo attempted to take by surprise
the palenque El Espinal, which was not far from Matanzas.[56]

The life of a ranchador was not about making political decisions in safe
and luxurious city halls but about food and water shortages, about blood—
maroons' blood and sometimes also their own—and about different kinds
and degrees of brutality. It was not an easy life. For maroons, however, life in
the mountains was even more difficult.

Maroons

In making the decision to flee to the mountains, plantation slaves were well
aware of the countless difficulties that they would face. Days and nights exposed to rain and wind and serious problems finding food were the most
likely immediate consequences of any escape. Some lucky escapees found
caves in which to hide from the elements, but others did not; all anticipated
the possibility of fights with rancheadores and their ferocious dogs, capture
and punishment, and even death. For those who managed to enter maroon
communities, known as palenques or rancherías (which were smaller and
less protected maroon villages), life was often easier, or at least more organized, though rancheadores and their dogs remained a continual danger.
Despite the obstacles, however, slaves continued to escape. Throughout the

nineteenth century, runaways constituted a regular component of the already complicated Cuban rural society.[57]

Most runaway slaves were what officials called cimarronaje simple. Thousands of cases of escape documented in Cuban archives ended after one or two nights. Often, the runaway slaves did not even leave their estates; they simply hid in the coffee or sugar fields until they decided on their own to return or were found by the rancheadores. Some runaways stayed in the mountains, swamps, or forests for long periods of time, barely managing to survive, before they finally gave up and surrendered to the authorities or to their owners. In 1841, the captain of the jurisdiction of Cimarrones wrote Captain-General Jerónimo Valdés about this phenomenon: "In this captaincy under my command from time to time black maroons appear, some of them very skinny, and some others very old, giving the impression of being very sick, either from the bad treatments they receive or because they end up this way due to the lack of provisions they endure while living in the mountains."[58] In a similar vein, the civil deputy of the village of Remedios, Antonio María de la Torre, wrote early in February 1831 to the local council about a couple of runaway slaves who had appeared in his office the day before: "These slaves have arrived naked, forcing me to supply them with new clothes to cover their flesh; and after inquiring why they had come in such a state, the man said that they have been without any sort of clothes for at least two years, which was the time they had been missing from their owner's house, remarking that they would not go back because of the enormous punishment they had received there; and they begged me to give them another master who, in accordance with common human behavior, would not force them to make the decision to run and hide again."[59]

Because conditions in the Cuban mountains were so harsh, most maroons did not remain isolated.[60] Rather, they joined palenques and rancherías as a way of coping with their new life. Palenques fostered the growth of new communities, which were necessary for survival in a hostile environment full of dangers and unpleasant surprises. The highest expression of slave resistance, palenques were the only openly victorious outcome of slave resistance recognized by both slaves and owners.

Palenques were often wisely sheltered, defended, and organized. They were located inside caves, in the woods, in swamps, and on mountains. Some palenques were so well hidden that rancheadores frequently passed by without noticing them.[61] Their remote locations allowed maroons periodically to abandon and reoccupy them, moving from one palenque to another to avoid detection. Rancheadores unsuccessfully tried to prevent this from happening by burning everything when they uncovered a palenque. On March 28, 1815, rancheador Matías Pérez Sánchez found an old palenque on the top of a cliff and, after destroying the remnants of its huts, commented in his diary that he was "horrified because of the location of the place." He continued, "I could not do less than burn it down."[62]

In addition to choosing locations that were hard to find and attack, maroons also created fortifications and dug pits from which they could ambush the men who hunted them. Often, they also built other, more inaccessible refuges for use in emergencies. Lieutenant Manuel de Chenard, who was in charge of the troops who supposedly destroyed all the existing palenques in 1816, offered these conclusions about the maroons' fortifications and defenses:

> When they are followed by a group [of rancheadores], aiming at not finding themselves absolutely deprived of a place to hide, they choose other places no less isolated and with the most difficult ways of access, where they also grow some vegetables and fruits. . . . They also take precautions not to open footpaths to these refuges . . . or [else] to open some fake ones, seeded with sharp stakes of palo de cuava, surrounded by a horizontal incision, in a way that once it is nailed to the feet it becomes impossible to remove without surgical tools. And they also lay these traps in the paths they have prepared for their escapes. . . . Moreover, when the palenque is situated on the top of a mountain, they bolster their defenses with the use of huge rocks, which they throw downhill at the moment they believe appropriate.[63]

The defense of palenques was not restricted to choosing an adequate location and preparing traps for the rancheadores. Maroons had to avoid being found by the dogs or by vultures—which were often nearby—in order to survive.[64] Knowledge of warfare, often acquired in Africa, was also extremely important to maroons' survival. Maroons used every kind of weapon they could gather. Firearms were frequently part of their arsenal, as were lead pellets, sharp knives, machetes, and other types of weapons made of iron.[65]

Sorcery also constituted a valuable form of protection. On several occasions, rancheadores reported finding items related to "witchcraft" in abandoned palenques and rancherías. In 1829, Colonel Francisco Chappottín mentioned in a letter to Captain-General Vives the many dolls he had found in the palenque of Monte Escopeta.[66] Four years later, in 1833, Fernando de Osuna wrote that his men had recovered "an infinity of bags containing dolls adored by the blacks" in another palenque.[67]

Even water and food could become lethal weapons. Manuel de Chenard, an officer with vast experience hunting maroons, noted in his *Ynstrucción* of 1816: "Once arrived at a palenque, precautions must be taken to not touch any stored water, corn flour, or preserved meat found there, until certain that they are not mixed with some herbal compositions or deadly plant peels known and used by the blacks who grow old in the forests."[68]

Indeed, the defense of palenques depended to a large extent on the maroons' warfare skills and weaponry. The defense of the palenque also required leadership. Although palenques and rancherías did not always have defined leaders, maroon captains were often present.[69] The names of some important maroon leaders became famous among slaves in the plantation regions. Yará Gangá, Domingo Macuá, Pancho el Marqués, and José Dolores Congo are only some of the maroon leaders who gave hope to the slaves on Cuban sugar and coffee plantations. Obedience and commitment to their captains were often powerful motives for slaves to rebel or run away to the mountains. Some maroon captains fought until killed or captured and became legends among their companions. This was the case of José Ramón Mandinga, the one-handed slave who participated in the revolt of Guamacaro in June 1825, was sentenced to death some months later, and still managed to escape and remain free for three years. During this time, he became famous among the slaves of the plantation area around Matanzas, who believed he could escape from any ambush thanks to his knowledge of the magical arts.[70]

A few captains apparently betrayed their own men, becoming spies or guides for the rancheadores. In January 1831, maroon captain Agustín Mandinga, also known as "Madre de Agua," surrendered after putting up a brave resistance to the dogs of rancheador José Pérez Sánchez. Agustín had been missing from his plantation for more than six years, during which time his reputation had grown among the slaves on the plains. Immediately after being captured, he offered to guide the rancheador to the refuge of his former subordinates. Whether he did so or not is unknown. The relevant pages in Pérez Sánchez's journal are missing, and no subsequent references to

Agustín appear in the journal.[71] In another case, the slave Manuel, presumably another captain of maroons, gained his freedom in 1829 as a reward for his services as a guide to the rancheadores. His letter of freedom was bought by the Royal Council of Agriculture and Commerce under the condition that he would continue to act as a guide for an undefined period of time.[72]

But the daily struggle for survival was also tremendously important. Maroons frequently tried to create communities in the mountains.[73] They created their own families or brought their relatives to live with them in the palenques. They built huts and, thanks to trade with plantations, managed to obtain a wide range of commodities, from agricultural tools to kitchen implements. After his capture, Francisco Mina recalled how one of his friends secretly visited the sugar mill of Don Miguel Herrera on the orders of his captain, Hilario Mandinga, in order to abduct the slave Rosalía Lucumí and bring her to be "his woman."[74] Rita Bibí, a former maroon from the palenque El Cedro, also testified about everyday family life in the palenques. She stated that "there were more than thirty huts and in each there was a maroon with his wife."[75] In some cases, there were apalencados who grew up in the freedom of the mountains. Tomasa Criolla, a sixteen-year-old maroon girl, was captured in 1835 after spending her entire life in the mountains; her mother had been a slave at the sugar mill San Francisco Alfaro until she had escaped almost two decades earlier.[76]

Internal rules played a significant role in the survival of every palenque. Very often, there were prohibitions on straying too far from the palenque alone or even accompanied. Lying or gossiping excessively were also punishable crimes. In El Cedro, every time a maroon lied or defamed any of his fellows the captain of the palenque ordered a number of lashes as punishment.[77]

Entertainment, pleasure, and religion were not absent from maroon communities. The palenques provided a good environment for sexual relations, and lively conversations in the freedom of the mountains were fueled by the rum smuggled from plantations and the tobacco that slaves dried and rolled by themselves.[78] Drumming festivities, whether profane or sacred, were also common. In some cases, as in El Cedro, maroons reserved a hut for religious gatherings. This hut, known as the casa de cabildo, was also used as a meeting point in the mornings for those who were going to the plantations to steal food and tools.[79]

Agriculture was necessary for survival, and much in the same way that slaves cultivated crops in Africa or on the plantations, they raised yams, beans, plantains, and other foods in the wild.[80] Maroons hunted wild hogs and various kinds of rodents (jutías) for meat to supplement their diets. Men were mostly occupied in hunting and in the defense of the palenques, while women were in charge of agricultural and domestic tasks. When the palenque was attacked, however, men and women often faced the enemy together, and there are reported cases of women being in charge of the palenques' observation posts.

Maroons' relationships with their companions on the plantations was vital to their survival. The clandestine trade with plantation slaves supplied the palenques with tools and all sorts of needed artifacts for their daily existence. Captain Manuel de Chenard provided a short explanation of how one such exchange worked: "These [the maroons] ordinarily bring the wax to the sugar cane fields, and from there the domésticos of the plantations export it to the city on bank holidays, selling it to Catalan merchants who provide them in exchange with hatchets, machetes, gunpowder, flints, plaits, trousers, salt, and other articles that these blacks take back to the place of deposit, where the maroons come to collect them."[81] Contemporary sources make it clear the maroons and the mansos (that is, slaves who had not escaped or rebelled) were in constant communication. Francisco Mina, for example, was captured while trading beeswax with his companions on the coffee plantation Landot.[82] In July 1820, ranchador Gaspar Antonio Rodríguez complained about the communications between maroons and mansos.[83] In October 1828, José Pérez Sánchez echoed Rodríguez's complaints, writing in his journal that because of "the refuge they usually find in the estates, dispersing themselves," maroons were difficult to find.[84] The most famous of all Cuban rancheadores, Francisco Estévez, also grumbled about this in 1841, stating: "This exceedingly grueling and expensive excursion would never be enough to exterminate the maroons of these regions, because as soon as they have notice of our arrival, they find refuge in the plantations, where they are well covered by the mansos."[85]

Surprisingly, maroons were not only protected by their comrades but often also by the owners and employees of the estates they visited. In August 1820, Rodríguez had to investigate and denounce the mayoral José López

Toledo and his son, who were not only sheltering maroons but also employing them for various works.[86] Referring to this situation in 1822, Rodríguez asked Captain-General Mahy, "How is possible, dear Sr., that I can conclude such a tough commission upon which lies the calmness of the society, if the same owners ignore their interests?"[87]

Maroons not only traded with mansos but also stole supplies in sudden after-dusk raids. More feared than day-to-day communications, raids were an effective way for maroons to obtain food and other supplies for the palenques, including new recruits. Reports of killed cattle or plundered fields were not uncommon in the plantation regions.[88] The most feared maroon raids were those that were intended to rescue their slave companions. In February 1832, the governor of Matanzas received word that maroons had attacked the sugar mill Jesús María in Sabanilla del Encomendador with the aim of freeing a slave who was in prison there after having being punished the day before. The letter recounted that the maroons had arrived after midnight, broken into the prison, freed their companion, and escaped to the forests.[89] In a similar case in 1831, a group of maroons attacked the sugar mill San José, west of Havana, sacking the huts and taking two slaves away with them.[90] In 1836, the mayoral of the coffee plantation San Carlos in Cayajabos complained about the "black maroons who cheekily and frequently come to the plains looking for provisions."[91] Small companies of the Spanish army and groups of rancheadores also occasionally suffered these raids. In July 1832, for instance, after facing strong resistance in their attack on a palenque near Canasí, Lieutenant Joaquín Figueredo and his men were surprised when the maroons counterattacked, killing all his dogs, injuring one of his men, and releasing the only two slaves who had been captured earlier that day.[92]

Communication and collaboration between maroons and mansos was exceptionally important for the protection of the palenques. Rancheadores frequently enlisted slaves from the plantations to infiltrate maroon communities. Often, this tactic elicited information that enabled the rancheadores to persuade the maroons to surrender or to identify the location of their palenques. In 1819, José Garcilaso de la Vega, captain of the jurisdiction of Cayajabos, informed his captain-general of the names of the most important maroon leaders in the region under his supervision. This information, he specified, had been obtained by some slaves who had been sent by the superintendent, José Antonio Ramos, and some other planters of the area to infiltrate the palenques.[93] In 1830, using the same tactic, José Pérez Sánchez managed to persuade fifteen of eighteen runaway slaves from the sugar mill La Luisa to surrender.[94]

To avoid attacks, maroons had to be able to identify and control the emissaries and spies sent by owners and rancheadores. Relying on their communication networks with their companions on the plantations, maroon slaves developed a counter-espionage system to alert them to informers. In February 1822, Gaspar Antonio Rodríguez suffered a massive setback after trying to infiltrate a nearby palenque with one of the slaves from the sugar mill San Roque. When his informer did not return on time, Rodríguez went after him in the mountains. Approximately a week later, he was informed that the slave had returned only after miraculously escaping from the maroons he was supposed to infiltrate. Rodríguez related that when his informer arrived in the palenque, the maroons were waiting to tie him up. They knew why he was there and even at what time he had been ordered to arrive. When Rodríguez asked the mayoral how this information had been leaked to the palenque, the mayoral replied that he believed that all three hundred slaves were responsible for passing information to the maroons.[95] In any case, when Rodríguez and his troops arrived in the palenque, they found nothing but empty huts and a few old provisions.

The Impact of Cuban Maroons

Cuban maroon communities—especially those in the eastern part of the island—have been the subject of some controversial studies. These works have suggested that the "palenques of Cuba," as they were known, were a real headache for nineteenth-century colonial authorities. Some palenques did indeed grow beyond the expectations of authorities and slave owners. The mountainous and scarcely populated area of Baracoa and its surroundings, which were virtually isolated from the rest of the urban settlements of the region, became the most likely place for maroons to settle and develop large communities of runaway slaves. The palenque El Frijol reportedly hosted around three hundred maroons between 1818 and 1819.[96] Yet even if the maroons of Baracoa did pose a serious threat in the late 1810s, their achievements were few, and they remained isolated from the rest of the island. Other regions also hosted maroon communities. The plantation area around Matanzas and Cárdenas hosted at least one famous palenque, El

Espinal, and some well-known maroon captains plundered and devastated the area throughout most of the first half of the nineteenth century. Nonetheless, the mountains and hills located west of Havana were the best-known refuge for runaway slaves.

Maroons feared surprise attacks, but once the fight started, they usually were ready to die defending their freedom rather than return to a life of slavery on the plantations. Maroons were heroes among plantation slaves. They embodied and symbolized slaves' undeniable and overwhelming desire for freedom and, on some occasions, were contacted to help in slave plots, as in the case of José Dolores Congo in the surroundings of Matanzas in 1844.[97] Relationships between maroons and mansos were always based on the defined limits of at least temporarily free and enslaved men and women. Despite the admiration that mansos felt for their runaway companions, they were aware of the tremendous dangers and privations that maroons faced day after day in the mountains. No wonder, then, that so many slaves refused to escape even after being liberated by groups of rebels.

The real magnitude of marronage in colonial Cuba requires further study. Although maroons were a constant presence in the island's countryside, their impact on society and politics seems to have been limited, at least when compared to that of their equals in neighboring countries.[98] Their daily life also needs to be carefully assessed in order to incorporate it into the vast body of knowledge that we already have about marronage in other areas of the New World.

Claudio Martínez de Pinillos, count of Villanueva,
who took a serious interest in the issue of slave
suicides in the mid-1840s

Catholic priest Juan Bernardo O'Gavan became the
most fervent champion of the slave traders' cause
before the Spanish Cortes

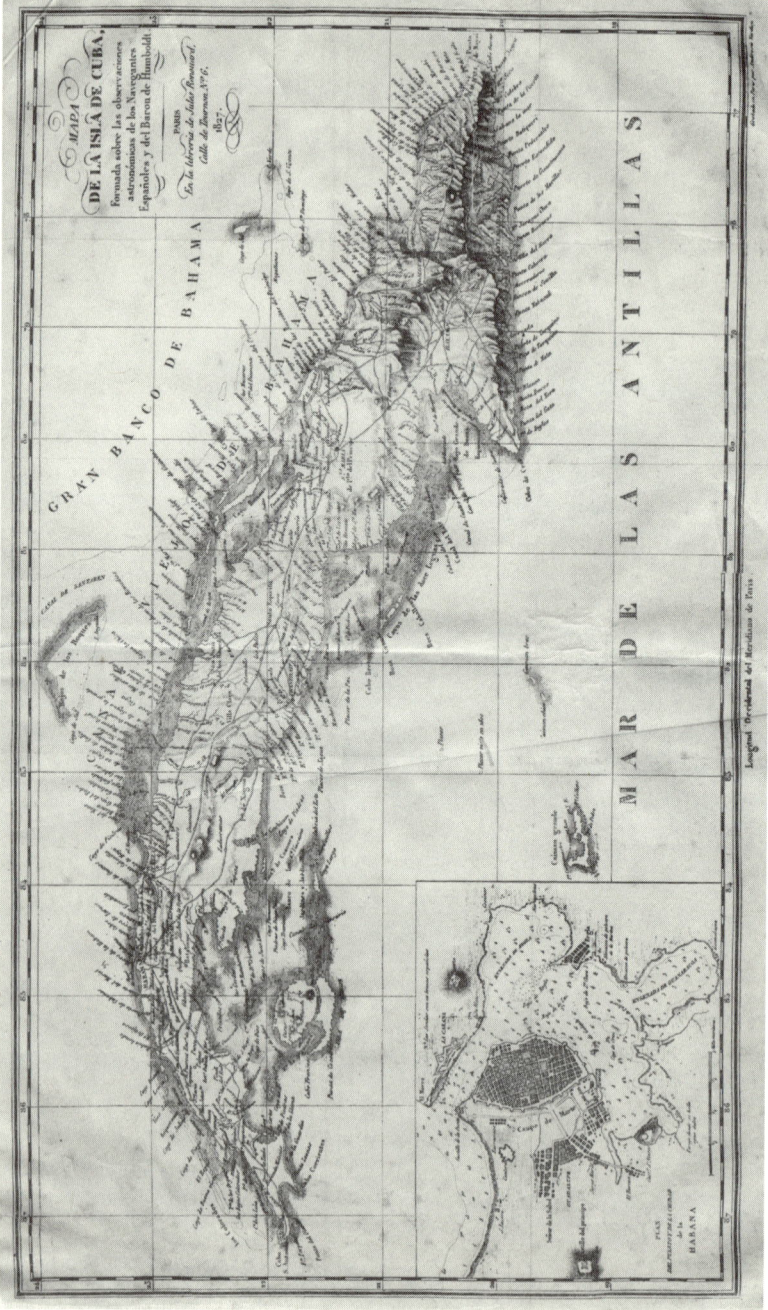

The island of Cuba in the nineteenth century

Slaves at the ingenio Intrépido on their way to work. It is not difficult to imagine them socializing and talking among themselves—perhaps in their native African languages.

Slaves working at the ingenio Amistad under the whip of a slave driver

Slave women washing their clothes and chatting at the ingenio El Narciso. It was within this sort of environment that slaves reproduced their hidden transcripts.

Slaves working the fields at the ingenio Santa Teresa á Agüica

4

❦

Suicides

A cucumber is bitter—throw it away.
　—Marcus Aurelius, Meditations, vol. 8.

Slave Suicide before and during the Middle Passage

Suicide was not rare in West and West-Central Africa.[1] In fact, very much in the Stoic tradition, it was a way out for the humiliated, the disgraced, and those with incurable illnesses. In an important episode in the history of the Kingdom of Oyo, the Oba Asamu, also known as Arogangan, killed himself after being asked to do so by a crowd. Before committing suicide, Asamu went to the outskirts of the town and from there cursed its future inhabitants. He then returned to his palace and—in a manner reminiscent of Cato's suicide in ancient Rome—refused to listen to the advice of his servants who urged him to fight, sent his relatives away, and took his own life.[2] It is important to mention here that according to legend the brave Shangó, god of lightning and thunder, ended his life by committing suicide.[3] In other West African regions, for example in Dahomey, suicide was not socially acceptable behavior and those who committed suicide were punished by "throwing their corpses to the beasts" and thus depriving them in this way of a proper burial.[4]

Before and during the Middle Passage, captured slaves committed suicide frequently enough to cause a tremendous problem for their captors. Just

before boarding the slave ships on the African coast, captured Africans took
their own lives in variety of ways. Captured slaves could repeatedly attempt
suicide until they were successful, or they could resort to other violent means
of resistance, such as self-mutilation or the assassination of crew members
and land-based merchants.[5] Slave traders' fear of such events was so serious that some contemporary writers, such as the Frenchman Jacques Savary,
urged captains to depart rapidly after loading their ships. Savary explained,
"The slaves have such a great love for their land that they despair to see that
they are leaving it forever, and they die from sadness. I have heard merchants
who participated in this commerce affirm that more Negroes die before leaving port than during the voyage. Some throw themselves into the sea and
others knock their heads against the ship; some hold their breath until they
suffocate and others starve themselves."[6]

And indeed, Savary was sadly right: many such incidents were recorded.
In 1812, for example, Captain Felipe Nery wrote that while the ship under his
command was entering the River Zaire in West-Central Africa, three of the
slaves he was carrying "[threw] themselves into the sea" after being whipped.[7]
Surviving slave testimonies are sometimes terrifying and almost unbelievable. Multiple reports attest to the general fear among the enslaved that their
white captors would devour their flesh or suck their blood. These fears led
many of them to commit suicide.[8] A remarkable case from 1737 illustrates
the strength of these assumptions. Just after the slave ship *Prince of Orange*
docked on the island of St. Christopher, over one hundred of the African
slaves on board jumped into the sea in a collective suicide attempt. A joke
made by a local slave had inspired their sudden determination to end their
lives: the slave stated that the newly arrived slaves' eyes would be put out
and eaten by their white masters. This idle jest led to the loss of thirty-three
human lives.[9] Beliefs in white cannibalism and white vampirism remained
alive throughout the history of the Atlantic slave trade. According to Moreau
de Saint-Méry, recently arrived slaves were especially afraid of their masters
after watching them drinking red wine, for they genuinely believed that their
captors were drinking blood.[10]

The punishments given to those who attempted to commit suicide
ranged from cutting their arms and legs to filling their mouths with boiling
lead. But the different measures taken by slave traders to prevent suicides
never succeeded in ending the attempts.[11] According to the log book of the

ship Hannibal, kept between 1693 and 1694, the "Negroes" who starved to death or who willfully drowned themselves did so in the belief that once dead they would "return home to their own country and friends again."[12] The bodies of slaves who had committed suicide were therefore mutilated by slavers, who cut off the corpses' arms and legs in order "to terrify the rest [of the slaves]," who were convinced that "if they lose a member, they cannot return home again."[13]

Once the slave ships were at sea, the situation did not tend to improve. Rather, the slavers' problems increased due to the slaves' isolation and the poor living conditions on board the ships. Some slave ships were packed with slaves in numbers almost unbelievable for the ships' size. Onboard punishments, compulsory dancing and singing, epidemics and other health problems, and inadequate food all made living conditions almost intolerable for the slaves.

The Spanish doctor Francisco Barrera y Domingo wrote in 1798 about the behavior of slaves in the Middle Passage, mentioning their different forms of self-destruction more than once in his text. Referring to "Viví, Carabalí, [and] Minas," he stated, "When they can't throw themselves into the sea, they get sad until they die." He added that it was sometimes possible to ease the slaves' pain by telling them lies about their immediate future. He warned, however, that should they realize the truth, they would wait for the right moment to jump into the sea, "because they believe that by doing this, they free themselves from the Europeans, and that they go back to their lands."[14]

Slave Suicide on Cuban Plantations

Slave suicides were a tremendous problem for slave owners from the antebellum South to the Caribbean and Brazil.[15] Slave owners and colonial authorities in Cuba always considered slave suicides to be pernicious and sinful acts, in accordance with the precepts of the Catholic Church, but their real beliefs doubtless stemmed more from economic realities than spiritual or humanitarian convictions. The acquisition of new slaves was difficult due to restrictions imposed by England on the Spanish slave trade in the treaties of 1817 and 1835.[16]

Authorities and planters in the New World believed that the imported Africans were absolutely convinced that by killing themselves they could return to their homes and families in Africa. This rationale appears in many official documents and in a wide variety of contemporary correspondence. The Cuban authorities' understanding of African slaves' cosmologies—and by extension their reasons for committing suicide—was quite poor, however, until well into the nineteenth century. Members of the Cuban plantocracy were often aware of the ideas of the French Enlightenment. They visited and sometimes studied in Europe and the United States. Many knew more than one language and were themselves foreigners or had foreign forbears. They discussed the writings of Rousseau, Montesquieu, Voltaire, and many other European philosophers, both publicly and privately. This was particularly the case among the civil authorities and intellectual leaders of the planter and merchant classes. Their approach to the issue of suicide, however, did not differ at all from that of other inhabitants of the island. Consequently, their understanding of suicide remained deeply biased by the dogmatic texts of the millenarian Roman Catholic Church.

The Catholic approach to the act of self-destruction was (and still is) surprisingly simple. Any suicide, regardless of its motivation, was seen as an evil act carried out against the will of God. Suicides had no right to repose in holy ground, and it was assumed that their souls would endlessly burn in hell. The inhabitants of the Catholic island of Cuba—slaves included—were taught that self-murder placed their souls out of range of salvation and divine forgiveness. Keeping in mind the official position of the Church and influenced to a large extent by it, authorities and planters dealt with slave suicides in their own peculiar ways. They used repressive measures, including the physical mutilation of suicides' corpses, to dissuade slaves from copying the actions of their dead companions. Havana's royal physician, José Antonio Bernal, a well-educated man, analyzed slaves' beliefs about suicide in a letter to the Military Commission written after three of his slaves had killed the mayoral of his sugar plantation. After demanding the highest penalty for the murderers, he begged, "We certainly know that slaves from the African coast persuade themselves in their ignorance that death at gallows opens the way to their homeland, releasing them from slavery. . . . Therefore, this punishment is inefficient if we want to teach the others a lesson. . . . They must be executed by [being shot in] the back as usual, then hanged and their heads mutilated, to be exhibited where it is convenient."[17]

Individuals of diverse origin and social status reaffirmed Bernal's opinion about the reasons slaves took their own lives and how Cuban slave owners should handle the problem. In 1790, for example, authorities and slave owners from Havana wrote to King Charles IV of Spain to complain about the recently approved Black Code of 1789.[18] In this letter, they accused their slaves of being uncivilized and intrinsically evil. The slaves' "Pythagorean beliefs"[19] in reincarnation was cited as the philosophical basis for slave suicide: "They are barbarous, daring, ungrateful of benefits. . . . Good treatment makes them insolent; their temperament is hard and rude; several of them never forget the error of the Pythagorean transmigration they learned in their early childhood. That is why they show little fear of committing homicide upon themselves."[20] Forty-one years later, when judging rebel slaves from the coffee plantation San José, experienced public prosecutor Francisco Seidel called for the death penalty for the slaves' three Lucumí leaders— Simón, Andrés, and Rafael. While the execution of slaves was a common event in nineteenth-century Cuba, Seidel, like Bernal, specified that the prisoners should be shot in the back "instead of hanging them as it is ruled, since as it has been demonstrated this kind of execution does not impose or scare these people like the first one."[21] Seidel also suggested mutilating and publicly exhibiting the head and right hand of Simón Lucumí, the principal rebel.[22] Twelve years later, Vicente Pérez, mayoral of the coffee plantation La Juanita, located a few miles west of Havana, testified before a public prosecutor after the slave Miguel Lucumí hanged himself from a tree. Shocked by the event, Pérez said that Miguel was an outstanding worker and that he never had had to flog him. Asked about the cause of Miguel's death, Pérez stated that the slaves "believe that when they die they go to their land" and that Miguel had committed suicide for this reason.[23]

It is quite noteworthy that over a span of more than fifty years, a medical doctor, members of Havana's city council, a public prosecutor, and a mayoral held such similar opinions about the question of why slaves took their own lives. These statements are the best proof of the widespread belief in reincarnation among African slaves. They also offer an excellent point of departure for an analysis of slave owners' and colonial authorities' understanding of this aspect of slave cosmology.[24] Authorities and planters discussed whether suicides among their slaves were provoked by religious, social, or medical reasons. It was clear to them that when slaves killed themselves they were trying to escape their condition of bondage. An unfair social system was the cause of slaves' suicides, which were the ultimate consequence of it.

Material evidence supports the idea that slaves who took their own lives literally believed they would journey back to Africa. In 1847, the count of Villanueva investigated why suicide rates among African slaves were rising. Relying on his own experience as a planter—he owned two important sugar mills west of Havana—he commented that slaves usually hanged themselves from trees or in their huts. "When they do that," he wrote, "they wear all their clothes, put in their hats the food that they did not eat, and sometimes even bring their animals to the place where they will die, in order to return well supplied to their native countries, where they believe their bodies and souls go."[25]

Several cases seem to confirm the count's explanation of why slaves provisioned themselves before committing suicide. In March 1825, Joaquín, a slave on the ranch Los Acostas in Guane, committed suicide by hanging. In this case, the authorities remarked that Joaquín was dressed "with new trousers and new shirt, and with pig leather shoes; he had a handkerchief tied to his head and in one of the pockets of his trousers a small bag with stuff to make fire; he also had a machete, a knife, and a hat made of palm leaves; everything was on the floor."[26] In June 1844, similarly, Dionisio Lucumí decided to hang himself from a guava tree. He wore his best clothes, and authorities found tobacco inside the hat he was wearing.[27] When Miguel Lucumí committed suicide, the mayoral, Vicente Pérez, noticed that Miguel was fully dressed in trousers and a shirt, with a handkerchief on his head and a machete and a hat made of chaff both lying on the ground near his body.[28] In contrast, however, some reports of slave suicides describe what the slaves wore only briefly or not at all. Detailing the suicide of Paulino Lucumí in June 1846, the official report mentioned only that Paulino was wearing his everyday clothes and that these were already well-used.[29]

Suicides were common enough to raise important questions for Cuban authorities and planters, such as which ethnic groups were most prone to self-destruction, what their religious beliefs were, how they took their own lives, and how to dispose properly of their bodies. The ethnic question was a perennial one. In 1798, as we have seen, Barrera y Domingo considered the Carabalís, Vivís, and Minas to be the ethnic groups most predisposed to commit suicide. From the 1820s onward, however, Lucumí slaves attracted most attention from the authorities, planters, and the general public.[30] The count of Villanueva, again relying on his own experience as a planter, opined in 1847 that in order to prevent slave suicides, it was essential to know which ethnic groups were the most prone to suicide: "By acquired experience in the government of the slaves in my holdings, I have confirmed the idea that among the different African nations or tribes to which the imported slaves belong, there are some which easily develop despair, uneasiness, mental disorders, and the other ordinary causes of suicides. Those who belong to other tribes rarely kill themselves."[31] Though the count did not refer to any particular African group, he was well aware of the preponderance of suicides among Lucumí and Carabalí slaves. Some of the biggest slave revolts of the period, led by Lucumí slaves, ended in collective suicides; the great uprisings of Guanajay in 1833 and Bemba and La Guanábana in 1843 are only three of the best-known examples. These rebellions, especially the last two, were still fresh in the count's memory. In his letter to Captain-General O'Donnell, he recalled how the scary atmosphere of the years 1843 and 1844 had led to a large number of suicides among "people of color," including slaves.[32]

The first recorded account of collective slave suicide in Cuba relates to a revolt that occurred on the coffee plantation Tentativa in January 1827. Two days after crushing the revolt, Colonel Joaquín de Miranda y Madariaga reported that up to eighteen slaves, most of them Lucumís, had been discovered hanging in the trees of the nearby forest.[33] This was also the first Lucumí-led revolt in the Cuban countryside. In the following years, every documented collective suicide among African slaves was led by Lucumís. Collective suicide, then, seems to have been a consequence of the increase in the importation of Lucumí slaves into Cuba after the fall of the Oyo Empire in 1820.

This ethnic theory has only one transparent weak point, namely, that the amount of time newcomers had to adapt to their new lives also determined the suicide rate to a remarkable extent. The count of Villanueva noted the importance of slaves' "seasoning" in discussing the causes of slave suicides, and planters were usually aware that it was necessary to give better treatment to recently arrived Africans. Such slaves were often assigned the easiest tasks on the plantations, and it was a common practice not to force them to work very hard during the first months due to slave owners' fear of rebellion and suicide. When these practices were not followed, the consequences could be dreadful.

In July 1835, the mayoral of the sugar mill La Magdalena committed a number of mistakes that led to a genuine disaster. First, he decided to send fourteen Lucumí slaves who had arrived on the plantation only three days earlier to work in the fields. Predictably, they decided to ignore his orders, but they were nonetheless forced to do their task. The next morning, two of them were found dead, hanging in a tree, and the rest again refused to go to work, "preferring to die instead." The mayoral, pushing the limits once again, sent them to the fields. Not content with this, he decided to line them up in front of the bodies of their dead companions. The testimony of Domingo Lucumí, one of the remaining slaves, offers the best description of what happened next: "We found many whites, who, separating the twelve of us who were left, aligned us in front of our dead companions. Seeing this, I told the remaining slaves that since our friends had died, we should die too, and therefore we attacked the whites with our machetes."[34] Domingo and two of the "many whites" he mentioned were seriously wounded. Ildefonso Lucumí threw himself in the river and drowned, while another rebel was found hanging from an avocado tree two days later. The last casualty was Serapio Lucumí, who died a couple of months later, presumably as a result of the wounds he received during the battle and subsequent punishment.

Frustration and rage incited these slaves to kill themselves and to fight against the white men who were trying to punish them. It was the unfortunate performance of the mayoral and his aides, however, that was the ultimate cause of the revolt. He ignored the origin of the slaves and the small amount of time they had been given to adapt to plantation life, and these two weighty mistakes provoked a collective act of resistance that involved both a rebellion and suicides.

Authorities and planters considered the means of slave suicide to be the key to understanding slaves' "savage" behavior. The count of Villanueva wrote in 1847 that slaves invariably hanged themselves; five decades earlier, Barrera y Domingo had reported that they most frequently drowned themselves in wells and rivers. But there were also many cases in which slaves stabbed themselves or slit their own throats. Sometimes, slaves chose to burn themselves to death, and a few even jumped into boiling sugar pans.[35] It was not unusual for maroons to jump to their deaths from cliffs; the most renowned slave hunters of the period left accounts of this practice. After dispersing the maroon slaves of a palenque in the mountains of Pinar del Río, José Pérez Sánchez noted that many of the slaves jumped from the cliffs without thinking twice about the danger. A quick search at the bottom of the cliff later revealed the body of a woman and bloodstains left by other slaves who managed to escape.[36] A year later, Pérez Sánchez referred back to these events when he wrote to the royal council, "It is likely that many of them [the maroons] have killed themselves jumping from the large rocks and cliffs."[37]

Some suicidal slaves made more than one attempt to achieve their goal. Trinidad, a slave from the sugar mill Jesús María, and Julián, a slave from the sugar mill La Cuchara, both tried repeatedly to escape and commit suicide.[38] Such attempts were sometimes taken to extremes, as in the case of the just-arrived Nicolás, the slave of Antonio Vidal, in Cayajabos, west of Havana. In May 1822, Nicolás learned that his attempt to escape had been frustrated and decided to cut his throat with a machete. Although the white plantation workers subdued him, he escaped again precisely when the situation appeared to be under control, and this time he managed to cut his penis. Although Nicolás did not die, his condition was extremely critical for months, and presumably he was never again forced to go to work in the fields.[39]

Religion played a fundamental role in the authorities' decisions about how to cope with the bodies of suicides. Villanueva briefly mentioned that in order to persuade slaves to stop killing themselves, some plantation overseers resorted to burning the bodies of suicides.[40] He recognized, however, that this method did not have very good results.[41] The bodies of suicides were excluded from the consecrated ground of Catholic cemeteries. In Havana during the first half of the nineteenth century, suicides were, curiously, buried in the potters' field for foreign Protestants known as the cemetery of the Americans or the Englishmen.[42]

Suicides among African slaves were thought to be the direct consequence of their savagery and lack of religious education.[43] In 1838, the defender of the rebel slaves of the coffee plantation Clarita, a lieutenant of the Galician regiment, declared, "My defendants were not men who always keep reason in its place; their ignorance is so great that most of the time they cannot understand half of what they hear; they have the appearance of men, and the rest is pure irrationality."[44] Throughout the nineteenth century, slaves' attorneys frequently made similar arguments. Beyond any doubt, elements of racism permeated the discourse of Spanish officers and of the white population in general. These arguments were often quite effective. Very often, slaves' lack of understanding of Spanish language and laws, and even their physical appearance, was entered into the record as evidence of their lack of culpability.

In light of the slaves' purported savagery, Cuban authorities and planters—many of them literate people with a solid knowledge of the works of the

Enlightenment—defended their right to perpetuate the slave system for humanitarian reasons. At the beginning of the nineteenth century, most Cubans were convinced that by bringing Africans to the Americas they were transforming them into civilized persons, an improvement on their savage origins. The slave trade became synonymous with progress and civilization in the discourse of the Cuban intellectual, merchant, and planter elite. The behavior of recently arrived slaves, particularly their violent acts of resistance, seemed to reaffirm the imperative to help them become Christian human beings. Newspapers and official documents supported this position throughout the first half of the nineteenth century. Among the most abominable transgressions of the savage Africans mentioned by Cuban elites was their propensity to commit suicide.

In 1821, Cuban priest Juan Bernardo O'Gavan went to Madrid as one of the island's deputies to the recently reinstalled Spanish Constitutional Assembly. His mission was to promote the restoration of the transatlantic slave trade, which had been legally abolished by the Treaty of 1817 between Spain and Great Britain. O'Gavan told the assembly that his aim was not to make "an apology for slavery," but his speech was the closest imaginable thing to it. Showing off his intellect, O'Gavan catalogued a series of climatic, geographic, epidemiological, and economic factors that contributed to the improvement of Africans' lives once they were transported to the New World. He called the peoples of central Africa "idle and useless" and stressed their misery, disorder, and stupidity.[45] In closing his argument about the differences between savage Africa and the civilized Americas, he described plantation slavery in preposterously idyllic terms:

> The black man lives surrounded by his family, with his wife and children, in his house and in his fields. He has the freedom to go fishing and hunting; and when he is hard-working and has some talent, he enjoys some comfort and even a certain degree of luxury. When he falls sick, he is treated with great care; when he is old, far from the problems of his own subsistence and his family's, he is treated with generosity. When he gains the necessary money for his deliverance, he becomes a freeman in disposition of all his faculties. Our special laws highly favor the good treatment and the freedom of the blacks, opening for them all the opportune roads. To conclude, these men, who would be indomitable wild beasts in Africa, learn and practice among us the precepts of the religion of peace, love, and sweetness, and become part of the great evangelical society.[46]

In his shameless speech, this Roman Catholic priest—one of the most influential figures in Cuba of his day—ignored what he knew was really happening just a few miles away from Havana.[47] He did not mention the massacre
that followed the discovery of the conspiracy of Aponte in 1812. He intentionally overlooked the hundreds of ships loaded with new victims that were
arriving every year on Cuban shores. What is more grievous, by defending the
slave system in such a blatant way, he voluntarily abandoned the thousands
of souls that were daily whipped and shackled on his beloved island.

In contrast, when asked by the bishop of Havana and the captain-general
of Havana how to account for the frequent suicides among African slaves,
the count of Villanueva offered a more honest response: "More than the fa-
naticism and more than the innate characteristics of African slaves, their
state of servitude should be considered as the main cause of their suicides.
No matter how easy this state can be and even when their destiny can look
preferable to the scarcity and uncertainty of European laborers, the freedom of the laborers still is and will always be preferable."[48] Many planters
and members of the ruling group privately shared the count's opinion. They
were less able to express this opinion than Villanueva, however, whose almost limitless power enabled him to speak freely.[49]

Most frequently, slave owners relied on religious instruction in their attempt to prevent slave suicide. Local authorities and planters had few doubts
that slave owners' failure to Christianize their slaves played a significant role
in the increasing number of slave suicides. The marquis of Arcos, a well-
known and rich slave owner, wrote in 1842 that fulfillment of religious precepts was the only way to prevent the excesses of the slaves.[50] The count of Fernandina went further, stating that the abandonment of religious instruction
among the slaves was regrettable. He also lamented the fact that slaves were
considered by their owners to be animals without the right to the afterlife.[51]

Discussions of this matter grew so intense that Captain-General Leopoldo
O'Donnell, answering the complaints of the bishop of Havana, ordered an
investigation into the causes of suicide among the African-born inhabitants
of the island. For this task, he appointed the versatile and experienced prosecutor Ignacio González Olivares, who made a detailed investigation of slave
suicides between April 1839 and November 1846. González Olivares drew his
figures not only from slaves but also, for purpose of comparison, from the
rest of the Cuban population. He discovered that 1,337 suicides took place

during his studied timespan. Of this number, 1,171 suicides (87.4 percent) were committed by slaves.[52] González Olivares concluded that frequent floggings and slaves' lack of religious instruction were the two principal causes of slave suicides. Echoing his contemporaries, he stated that slaves preferred to die than to work because "they have no idea of the next life, or they have it wrong, obscured by idolatry."[53]

In his judgment about slave suicides, the count of Villanueva was extremely clear and explicit: "Religious education is not and cannot be a fast work; and it cannot be assumed that those who have decided to kill themselves will desist from their intention by other means than by an extreme vigilance.... Good treatment, a rational working schedule, the welcome they receive from the old slaves, and above all the constant care of the employees on the holdings are the principal means to put off African slaves from the idea of committing suicide."[54] Before these lines were penned, never had anyone in Cuba written about suicide among African slaves with such understanding. Villanueva's experience as a planter and his life service as a bureaucrat to the Spanish crown gave him the experience to tackle the issue in a serious way. Even the bloodthirsty Captain-General O'Donnell was forced to recognize that the count's analysis was right on target.

Nevertheless, after 1847, slave suicides continued to be a serious problem for slave owners and authorities. The result of the exchange of letters among Captain-General O'Donnell, the bishop of Havana, the count of Villanueva, and the prosecutor González Olivares was an official decree aimed at improving religious instruction among the slaves. Nothing was ordered to ease the harsh conditions of work in the cane fields or to limit the quality and quantity of the punishments being administered.

Conclusion

The count of Villanueva's profound critique of the conditions of slavery and the causes of the rise in slave suicides resulted in nothing. It seems that all the knowledge acquired by the illustrious members of the government and the planter elite, all their reading of Rousseau, Voltaire, and Montesquieu, was useless when it came to easing the lives of their slaves. They did not wish to change the oppressive form of slavery practiced on the island, despite the fact that they knew very well that it was—as Villanueva rightly wrote—the main reason for the frequency of slave suicide.

On Cuban plantations, slaves committed the desperate act of suicide in answer to the unfair society to which they were brought to live. Torn from their beloved relatives, friends, and the country they knew, they resorted to suicide as a viable way to escape. On some occasions, they killed themselves for individual, selfish reasons, since integration into new plantations was sometimes very difficult and could be aggravated by the behavior of overseers and masters.[55] In the collective suicides that followed some revolts, slaves killed themselves for altruistic reasons, since they had tried to lead their friends to a better life, either by crushing their oppressors or, in defeat, by going back to their homelands. The excess of social regulations also provoked both premeditated and spontaneous suicide among slaves: consider the example of slaves who killed themselves after being punished by their masters or disappointed by the ineffectiveness of the colonial laws.

In sum, then, the rate of suicide among the slave population of Cuban plantations seems to have become a serious social issue in the first half of the nineteenth century for a number of reasons. Factors such as the ethnic origin of slaves, the amount of time they had spent in Cuba, their cultural and social backgrounds, and their living conditions largely determined the frequency with which they committed suicide. No doubt, slave suicides—whether or not they were performed as acts of resistance—were a constant matter of concern for Cuban authorities and slave owners. Every slave suicide represented an economic loss as well as a signal to the outside world that Cuban slavery was not the humane system that many slave owners tried to portray.

5

Slaves' Use of the Colonial Legal Framework

The repression and honorable punishment of crimes are essential elements for the preservation of nations.

—The count of Villanueva, July 26, 1830

On a mid-July day in 1843, José María Lucumí walked to the house of the governor of New Filipina—today Pinar del Río—with the sole idea of denouncing the mistreatment that he and his slave comrades had been enduring day after day. José María told the governor, Carmelo Martínez y García, that his master had repeatedly denied him his right to find a new master. He also informed the governor that the slaves on his master's estate were forced to work every day of the week and that his master had stolen and sold their pigs, thus depriving them of their only legal source of income.[1]

Martínez y García apparently felt some kind of sympathy for the slave, because he gave José María legal recourse. The governor decided to find out as much as possible about what was happening on José María's plantation. In his pursuit of truth, he interrogated many of José María's companions, who did not hesitate to testify against their cruel owner, Vicente Espinosa. Luciano Carabalí, for example, accused Espinosa of whipping slaves at his will, while Andrés Lalá denounced him for stealing the "little money" they were able to save over months of hard work.[2] As a result, Espinosa was fined twenty-five pesos and ordered to give better treatment to his slaves. In this respect, the daring action of José María was a great success. However, he was denied his request to find a new master and was sent back to Espinosa. Disappointed with the final outcome of the process, he killed himself on the morning of August 29.[3]

Criminal proceedings against cruel slave owners like Vicente Espinosa were common in colonial Cuba, particularly after the turn of the nineteenth century.[4] Some slaves, especially those already habituated to plantation life, learned to use the law in their favor and tried to claim the few rights to which they were entitled. In a few cases, they were rewarded with partial or total success. On most occasions, however, they were defeated and humiliated by the colonial authorities, who usually tilted the balance of justice in favor of the masters and overseers.

In this chapter, I look at colonial slave laws and how they were used by those slaves who were aware of the few advantages the laws offered them. I carefully selected the cases discussed in this chapter from among hundreds of other representative cases that are not as well documented.[5] Although I have tried to include an equal number of cases involving African-born and Creole slaves, the predominance of the latter is apparent, not only in this sample but also in the archival material that I had the opportunity to consult.

Slaves and Slave Law in the New World

From the beginning of the establishment of slave societies in the Americas, social norms were used to restrain the slaves imported from Africa and their descendants. The legal corpus regarding slavery remained disorganized and often outdated, however, throughout the first centuries of European domination. In general, authorities and slave owners treated slaves according to local or regional practices and rules, rather than following properly instituted laws.[6]

The first serious attempt to devise a comprehensive code of rules to govern slaves took place 169 years after the arrival of Columbus in the New World. The slave code for the English colony of Barbados, the Act for the Better Ordering and Governing of Negroes, was signed and proclaimed at the end of September 1661. It consisted of twenty-three clauses, in which measures were adopted to maintain the slaves under strict control.[7] Some of the act's clauses are illustrative of the daily life of slaves on Barbadian sugar plantations. Clause 5, for example, recommended that overseers search the "Negro houses for runaway Negroes," which suggests that seventeenth-century Barbados's maroons tended to find shelter on plantations, a common occurrence in the plantation societies of the New World. It is clear that "runaway Negroes" were the main target of the 1661 act, which gives the impression that authorities in Barbados were obsessed with the runaway slave population of the island.[8] The British were also concerned to regulate other forms of slave resistance, and they took care to explain the many ways in which resistance could be averted by masters, overseers, and their neighbors. The act addressed slaves' working hours, clothing and feeding, and punishment. Contemporary European (mostly British) legislation deeply influenced the slave code of 1661. Proclaimed a year after Charles II was crowned king of England, it reflected the unsettled character of British law and the resulting need for order and stability, which the Crown attempted to impose not only in Barbados but also in England.[9]

The Barbados slave code remained the only systematic attempt to institute slave laws for the next twenty-four years. Then, in March 1685, King Louis XIV of France issued a slave code for the French colonies in the New World. The Code Noir (Black Code), as it was popularly known, aimed to control the large number of slaves in the French West Indian colonies. The Code Noir stressed the importance of slaves' observance of Roman Catholic precepts of obedience. Several issues received judicial attention for the first time, including slave marriages, burial practices, access to weapons and alcoholic beverages, assemblies, alimentation, manumissions, and the slaves' position vis-à-vis colonial institutions and laws.[10]

In the following years, the Danish, Spanish, English, and Portuguese also reviewed their slave laws, which became central to the smooth functioning of the slave-based plantation economies of the colonies. Denmark, for example, published its slave code in 1733, which was a piece of draconian legislation. None of its regulations benefited the slaves, and many of them mandated punishments of a medieval variety. Captured maroon leaders, for instance, were ordered to be tortured with red-hot iron pincers and subsequently hanged. Ordinary runaway slaves were to lose a leg or an ear and then to receive a hundred and fifty lashes. The punishment for slaves who stole varied from being branded with a hot iron to a hundred and fifty lashes to torture by hot pincers followed by hanging. Pincers torture and hanging were also the punishments for slaves who dared to threaten or lift their hands against their masters. In some cases, the forms of punishment established by the Danish were even worse. For example, any slave who tried to poison his master was to be tortured with pincers and then broken on a wheel.[11]

Although practiced in many places throughout the Americas, the barbarities enshrined in the Danish code were not enacted elsewhere. Virginia and New York promulgated their own slave codes in 1705 and 1712 respectively, and although these codes reflected the interests of rich white slaveholders, their strictures were not as revolting as those of the Danish.[12] And while the Portuguese and the Dutch were known for the cruelties they exercised upon their slaves in Brazil and the Dutch West Indies, the enlightened reforms of the marquis of Pombal in eighteenth-century Brazil fostered the development of new ideas and considerations.[13] Particularly important to this change of attitude were the works of Jorge Benci and Manoel Ribeiro Rocha, both of whom contributed to the transformation of the discourse about the treatment of slaves with new theories about the reciprocal duties of slaves and their masters.[14] In the new climate of opinion, slaves were given some important rights and permitted to use the law to their own advantage, though some serious limitations on their recourse to law remained in place.[15] Ultimately, neither the Portuguese nor the Dutch promulgated official slave codes, although Brazil had various criminal codes and collections of laws—such as those of 1830 and 1871—in which slave legislation had a central role. Thus, their slave laws and correctional measures depended on the ministers and governors in power, who acted in accordance with the customs of their time.[16]

Ever since contemporary witnesses—many nineteenth-century travelers among them—described and compared the severity of the punishments administered to slaves across the Americas, the treatment of slaves has been a recurrent topic for comparative analysis. In recent years, scholars have assessed the treatment of slaves and the laws devised to control them, remarking on the harshness or leniency of laws and punishments in the different colonies of the New World.[17] They have paid special attention to slaves' rights to legal marriage, manumission, property ownership, and the ability to testify in judicial trials, because it was the disposition of these rights that ultimately defined whether slaves were recognized as human beings or as property. Adele Hast and others have argued that in the antebellum South, "legal marriage did not exist for the slaves," and have pointed out that after 1723 in Virginia and other southern states, "manumission was frowned upon . . . and forbidden by the law."[18]

The legal channels open to slaves in the antebellum South were particularly limited. Precisely because slaves were considered as property, they were not supposed to make use of the law. As Arnold A. Sio has argued, the "legal status of the slave developed exclusively in terms of property as the result of the demands of an emerging capitalism."[19] This slave-owning society was keen to affirm its control over the slaves by any means necessary. Therefore, the chances of slaves using the law for their own benefit were very slender. Their limited opportunities to offer legal testimony have attracted scholarly attention since Kenneth Stampp and Frank Tannenbaum addressed them some decades ago.[20] Recent studies discuss the limitations on slaves' legal status and the transformation of those limitations over time. Sio gives us, in a few phrases, a sketch of the legal status of slaves in the antebellum South: "Slaves were legally incapable of prosecution as accusers either on their own behalf or on behalf of others. As a general rule the evidence of slaves was not admissible in court, and when it was taken it was taken by torture, for it could not be received in any other form from slaves."[21] Given their status under the law, manumission by court order was a very remote possibility—practically speaking, an impossibility—for most slaves living in the Old South.

In an article published in 1994, Arthur L. Stinchcombe examined some variables that, according to his thesis, "shaped rates of manumission" across various slave systems.[22] These variables delimited the impact of the degrees of control exercised by the English, French, Danish, Dutch, and Spanish while also taking other considerations into account, such as the differences between African and Creole slaves, slaves from large and small holdings, and slaves living in cities and those living on plantations.[23] Although Stinchcombe focused his study mainly on the Caribbean islands, his argument holds true for the rest of the Americas. In fact, it has been reinforced by scholars whose work is centered mainly on Brazil. In her outstanding study of nineteenth-century Rio de Janeiro, Mary Karasch suggested that slave women were more likely than slave men to gain their freedom. Karasch also estimated that around 90 percent of Rio de Janeiro slaves were never legally manumitted.[24] More recently, James H. Sweet has revisited the subject of manumission in Rio de Janeiro and has endorsed most of Karasch's findings. Sweet took into account some of the variables offered by Stinchcombe and concluded that "slave masters were most likely to free those with whom they shared some bonds of affection."[25] He examined the segments of the slave population that were most likely to be manumitted and concluded that Creole slaves were manumitted more frequently than African slaves. Among the latter, slave women from West Africa had the better chance of manumission, while the odds of Central African men to be freed were "extraordinarily slim."[26]

The fact that Brazilian slaves made use of the law from at least the late eighteenth century is beyond question. The ways in which they did so, however, as well as the limitations of their legal personality and resulting prerogatives, are still being debated. Relying upon exceptional primary sources, Keila Grinberg has discussed how Brazilian slaves took advantage of the law throughout the nineteenth century.[27] Not surprisingly, her conclusions follow the path of Stuart Schwartz, Mary Karasch, Hebe Mattos, and James H. Sweet, echoing Stinchcombe's thesis regarding the variables that affected manumission.[28] Grinberg points to the significance of geography in the process of manumission, writing that "the proximity to urban centers facilitated access to information and to the individuals that could help with the setup of legal processes to obtain freedom."[29] Residence in a city such as Rio de Janeiro or Salvador could make a huge difference in a slave's chances of manumission. In Brazil and throughout the Americas, plantation slaves had a harder time gaining their freedom than their urban comrades. Grinberg also stresses another of Stinchcombe's significant variables, namely, a slave's place of birth. She argues that "the majority of the slaves that demanded their freedom had been born in Brazil" and "had some sort of relationship with their masters."[30] Once these cases were brought to court, Grinberg argues, the slaves "had real possibilities of obtaining their freedom."[31] She concludes that legal processes were an "effective mechanism that slaves could use to obtain their freedom against the will of their masters."[32]

The fact that slaves had the right to complain and even to fight for their freedom in nineteenth-century Brazil stands in sharp contrast to the extensive legal limitations imposed on slaves in the antebellum South. In the Spanish Caribbean, after the slave codes of 1784 and 1789 were passed, slaves had legal rights that they often exercised. Of all the Spanish colonies, Cuba would eventually have the largest slave population and the greatest tensions between authorities, masters, and slaves before the law.

Slave Law in Cuba

When, in the 1790s, Cuban planters and merchants began lobbying for a royal order that would allow them to import larger numbers of slaves onto the island, they compared the benign ways in which they treated their slaves with the cruel treatment for which the French and the English were notorious. Cuba, like the rest of the Spanish possessions in the New World, was then regulated by three law codes: King Alfonso X's Siete Partidas (issued between 1251 and 1265 for Castile, Asturias, and Leon), the Ordenanzas of Santo Domingo (1521), and the Ordenanzas of Cáceres (1574).[33] By the end of the eighteenth century, these antiquated laws could barely cope with the complicated Cuban slave society to which they were being applied. But the recently crowned Spanish king Charles IV and his Cuban governors faced serious opposition to a new law code from those they governed. Slave owners objected to the imposition of any new code for the governance of slaves because they feared two potentially devastating consequences. First, they did not want their godlike privileges to be terminated by a royal decree regulating what traditionally had been left almost entirely to the discretion of colonial authorities, masters, and overseers. Second, they were afraid that slaves would take advantage of any new laws and use them to claim previously nonexistent or badly defined rights.[34] Despite their objections, a slave code for the entire Spanish empire was issued in 1789. This code was a step forward in improving the conditions of slave life in Cuban cities and in the countryside. Although many of its provisions were not strictly observed and, in fact, the conditions of slave life in Cuba deteriorated in subsequent years, the Black Code of 1789 drew a theoretical line for the behavior of masters and slaves.[35] This line was interpreted and often trespassed, but it nonetheless established the rights and obligations of all members of Cuban slave society.[36]

Cuban whites regarded the code as a perverse piece of legislation that granted slaves too many rights. This opinion remained unchanged for decades. As late as 1879, the abolitionist thinker and activist Rafael María de Labra considered the code to be a very broad-minded document, recalling that it was totally suspended in three districts and asserting that its "spirit" was resisted everywhere.[37] However progressive the code of 1789 was in some of its measures, it did not abolish the death penalty or severe methods of torture as legitimate punishments for slaves. Labra, like many others before him, seems not to have noticed that even when the code favored slaves—guaranteeing them clothing, proper nourishment, and treatment for the sick and elderly—many of its other provisions severely restricted slaves' daily lives in numerous ways.[38]

The next slave code for the entire island of Cuba was issued in 1842 by Captain-General Jerónimo Valdés, though a local slave code had been put in place for the region of Matanzas in 1825 and a black code for the island of Puerto Rico in 1826.[39] Within a year after taking charge of the Cuban government, Valdés sent a questionnaire to some of the island's richest and most influential planters, asking several questions related to the way they treated the slaves on their estates.[40] He wanted to know whether improvements could be made for controlling the slaves, who had been rebelling more frequently than before in the preceding years. He also wanted to establish a minimum threshold of slave rights in order to legitimize the Cuban slave system in the eyes of the rest of the world.

The questionnaire was not well-received by many planters, who once again complained that a new set of regulations would entitle slaves to rights that they would later want to exercise. Sebastián de Lasa, one of the planters who answered the questionnaire, did not hesitate to let Valdés know that a new slave code was absolutely unnecessary and could have dangerous consequences. He wrote, "In my opinion, with the current laws, supported by individual interest, the good management and order of rural estates is sufficiently guaranteed." Fearing new slave rebellions if a new slave code was promulgated, he commented, "If the slaves get to understand that the government is taking care to issue new laws to improve their existence, I am firmly persuaded that our authority will be seriously damaged, and the consequences will be incalculable and fatal."[41]

The count of Fernandina echoed Lasa's comments. He confessed that he was convinced that any new disposition for the slaves—"even if . . . aimed to ameliorate their condition of existence"—would have disastrous consequences.[42] José Manuel Carrillo, Wenceslao de Villaurrutia, and Domingo Aldama also supported this position. Aldama reflected that "any direct intervention of the government that somehow might lead the slaves to believe that they have rights will increase their demands from the owners."[43] Carrillo warned Valdés about the alertness of the slaves, who were always ready to interpret in their favor any measure that could offer them some independence.[44]

The slave owners also feared that slaves might read more into the meaning of any new rights they were granted or guaranteed than the captain-
general intended, which could have the effect of escalating slave uprisings
rather than preventing them. Joaquín Muñoz Izaguirre, one of the most lib-
eral among the consulted planters, reminded Valdés how fast news spread
among slaves in the countryside. "They talk, make their comments, and more
than once it is possible to hear them using words with double meanings and
[making] very purposeful remarks," he warned.[45] Wenceslao de Villaurrutia
made an even stronger effort to convince Valdés of a direct link between
establishing new rules to govern the slaves and encouraging more slaves to
revolt: "Our slaves . . . believe that they have the right to be well treated, and
that the work they do is their biggest disgrace and a result of the coercion
we use. When they notice any relaxation in discipline, they never attribute
it to the goodwill of their masters or mayorales. . . . This explains why most
of the slave revolts that occur in our countryside are provoked by excessive
good treatment rather than by the harshness they receive."[46]

Certainly, Villaurrutia was defending a very conservative position, that of
the richest and most influential planters of his time—a position that was in
fact shared by most planters. Like Aldama, Lasa, and the count of Fernandina, he stressed the importance of keeping the authority of masters and
overseers intact, and he warned the government not to meddle in such a
delicate matter. He concluded his comments with a prophetic statement:
"From the day the slaves stop considering their immediate governors as a
discretionary authority . . . ; from the moment they realize that their food,
clothing, accommodation, . . . and even their work, which is the most interesting issue for them, are all accurately regulated and disposed beforehand
by a superior authority to that of their owners, the prestige of those will
immediately cease, and with it will also cease the few rules they are used to
respect."[47] Although not all the planters who answered Valdés's questionnaire
painted such a dramatic picture, most of them agreed that new regulations
would bring with them fresh problems in controlling their slaves. Even Juan
Montalvo and Joaquín Muñoz Izaguirre, the most open-minded planters
to answer the questionnaire, gave their tacit support to these assertions. A
year later, one of the richest Cuban planters, the count of Santovenia, wrote
to Valdés, "I consider that the storm is already above our heads, and it is

compulsory to exorcise it before it starts and buries us all."[48] The count of Santovenia was, in many ways, an exceptional case: he wanted to exorcise the storm not by repressing the slaves but by abolishing slavery.

The planters' premonitions and fears were proved right. Just after the promulgation of the Black Code in 1842, several slave uprisings broke out in the western part of the island, culminating in the discovery of an extensive conspiracy in December 1843. The subsequent repression of free colored people and slaves who were involved in the plot was the bloodiest episode in nineteenth-century Cuba until the first war of independence in 1868.

Slave Rights in Cuba

After 1789, slaves in Cuba had a number of privileges granted to them by slave codes or practiced as customary law. The three most significant privileges conceded to Cuban slaves were the right to marry, the right to purchase their own manumission, and the right to find a new owner when they were subjected to abuse and harsh punishments.

Slave marriages were socially accepted, officially recognized, and often encouraged by Spanish law from at least the late eighteenth century. In 1789, Cuban slave owners were ordered to promote marriages among their slaves. Slaves were given the right to wed not only other slaves on their estates but also slaves belonging to other plantations. Slave owners were obliged to accept these dual-plantation unions and subsequently to sell one of the married partners—usually the woman—to allow the spouses to be together.[49] The Code of Valdés, passed in 1842, strengthened slaves' right to matrimony. While articles 29, 30, and 31 essentially repeated the regulations of 1789, the code also forced the owners of the slave couple to keep the couple's children with their mother until the age of three years.[50]

Slaves, especially Creoles, were also aware of—and keen to take advantage of—their rights to manumission and coartación. Throughout Spanish and Portuguese America, the right to manumission enabled slaves to purchase their own freedom.[51] Coartación was a specific type of manumission by which slaves were allowed to purchase their freedom on a gradual basis, even against the will of their masters, provided that they presented at least fifty pesos to make a first payment.[52] As Rebecca Scott has described coartación, "Under Spanish law, a slave who made a substantial down payment on his or her purchase price—thus becoming *coartado*—gained certain privileges. He or she could not be sold for a price greater than the appraised value at the time of the *coartación* and was entitled to a portion of the rental if hired out."[53]

In order to finance their purchase, slaves depended on different sources of income. In the cities, slaves were frequently hired out to perform specialized tasks. Those who worked in sugar mills and coffee plantations were customarily allowed to breed and sell some small animals, including pigs, goats, and poultry.[54] Rural slaves were also entitled to harvest small plots of land, known as *conucos*, and to sell their products, in some cases to their own masters.[55] Frequent references to conucos appear in contemporary official documents. Most authorities and owners were happy to let their slaves breed animals and cultivate plots of land; indeed, some slave owners considered these concessions to be indulgences essential to keeping the peace among their slaves. Jacinto González Larrinaga, for example, boasted about his ability to preserve the peace on his sugar estates thanks to the existence of the conucos.[56] Domingo Aldama commented that once slaves owned some property, "their habits improve, they become cleaner, they do not fall into the temptation of running away, and they become enthusiastic helping themselves and their owners."[57]

Finally, Cuban slaves had the right to ask for a change of owner when they had strong reasons to complain about the treatment that they were receiving. This right was not universally known to slaves, particularly those recently arrived from Africa, nor was it regulated by any legal disposition. Instead, it was an important part of the customary law as it was practiced in Cuban cities and rural areas in the nineteenth century. To facilitate slaves' access to fair courts, the existing colonial institutions allowed them to secure the protection of any person who could represent them as a *padrino*, a term used to denominate a person—usually white—who would act on behalf of the slaves before their masters and local authorities. Spanish lawmakers also created the post of *síndico procurador* in each jurisdiction to look after the proper treatment of slaves and to serve them as a sort of attorney. The síndicos procuradores were "public servants who acted as protectors of the slaves and who were supposed to know and decide in first instance about their demands."[58]

To a certain extent under Cuban law, as in the rest of the Spanish colonies, slaves possessed a legal personality. This fact allowed them to exercise their limited rights in their own favor. Slave marriage was a common practice, perhaps due to its deeply religious significance within Spanish culture: urban and rural parish records from the sixteenth century onward demonstrate that slaves married and remarried without confronting many obstacles.[59]

Although manumission and coartación were mainly practiced in urban areas, some rural slaves also learned about them and began to approach local authorities to petition for their freedom with increasing frequency, especially after 1860. Studies on manumission and coartación in Cuba are not plentiful, but in recent years some important works have begun to reassess these slave rights by using cases documented by the colonial authorities. Unfortunately, these cases are overwhelmingly urban. Little is known about the use of manumission and coartación by rural slaves before 1860. The fact that slaves had the opportunity to complain and be heard and, more important, to ask to be placed with new owners proves that despite the rigors of the Spanish slave system, slaves had official and customary legal channels to claim their rights, raise their voices, and change the course of their lives.

Methods of Contention

The planters consulted by Valdés in 1842 were well aware that excessive punishments and abuses of slaves were common occurrences on estates. In the first half of the nineteenth century, physical punishment was commonly used to control slaves. The management of sugar mills and coffee plantations was usually entrusted to a few white employees, while administrators were in charge of keeping the books and investing profits. Mayordomos worked directly with the slaves and took care of the estate's premises. It seems to have been the case, however, that they were often forced to take care of the books and even to oversee every aspect of life on the estates.

Overseers were indispensable employees, and they were often described as the most controversial characters in plantation life. They were directly in charge of the slave labor force. They tolled the bells in the mornings, just before dawn, to awaken the slaves. They cracked the whip all day long in the fields and elsewhere on the estates before, during, and after working hours. They usually organized and controlled the distribution of food, and they held the keys to all the buildings. Ultimately, they were the most hated men in the Cuban countryside, with the sole exception, perhaps, of their alter-egos, the rancheadores, or slave hunters.

There are thousands of testimonies characterizing Cuban overseers as unpolished men with no respect for anything but money. As early as 1791, they were the target of a public attack by one of the oldest Cuban newspapers.[60] Juan Montalvo, one of the planters consulted by Valdés in 1842, expressed his concerns about the all-powerful overseers: "They are the ones who sacrifice the slaves. . . . Those men receive our estates and later on they have more authority there than the owners, because it is their unbreakable golden rule to always do the opposite of what the owner disposes, and never to tell the truth."[61]

But overseers did not carry out their work all by themselves. They were frequently assisted by slave drivers, commonly known as contramayorales. Though slave drivers could be whites or free people of color, they were usually slaves from the estate on which they worked. Slave drivers were ambivalent characters on Cuban plantations. In several recorded cases, they were as hated as overseers or owners, and thus it is not surprising that they were frequently victims of slave uprisings. Nevertheless, they were also regularly involved as active participants in plots and rebellions, often in a leadership capacity.[62]

Punishment was common on Cuban plantations. Planters and their employees often ignored the daily limit of twenty-five lashes per slave allowed by the slave codes of 1789 and 1842. Slaves could be put in shackles and forced to do field labor while carrying heavy chains. They could also be deprived of food and water or put in the stocks for days or weeks. Occasionally they could also be sent to prison for different reasons. In extreme cases they could suffer even harsher punishments, sometimes losing their lives in the process.[63] There are several accounts of the excesses of Cuban planters and employees. For example, as late as 1862, Esteban Santa Cruz de Oviedo, the planter who uncovered the conspiracy of La Escalera in 1843, gave 560 lashes to his slave Enrique, who died in consequence. Oviedo also branded the moribund slave with a hot iron.[64] Two years later, an even more barbarous series of tortures took place in Bejucal, a few miles south of Havana. According to several witnesses, the mayoral and the ox-driver of the sugar mill San Rafael perpetrated a large number of acts of violence against their slaves. The prosecutor assigned to conduct the criminal investigation recounted a horrendous and almost unbelievable story:

> The Negro Eduardo was chained, having a huge laceration in his right foot infected by worms . . . he was apprehended and after being beaten and whipped,

the ox-driver let his dogs bite [Eduardo's] private parts and as a result he was still peeing blood. . . . The Negro Diego was tied up to a tree and forced to bark continuously like a dog, being whipped every time he stopped. . . . The Negro Hipólito, who was already very sick, was pushed towards a nest of wasps while the ox-driver whipped the nest, exciting the wasps. . . .In front of everybody, Matilde was forced to sit in a chair where the mayoral, wearing his glasses, examined her private parts looking for worms while he shouted obscenities. . . . On another occasion, being in the fields, the mayoral ordered [some other slaves] to grasp Isabel; then he ordered the slave Hilaria to roll up her clothes and to urinate in her mouth; and since Hilaria refused to obey his order he whipped her, forcing her to urinate in the face and mouth of Isabel.[65]

Although both cases resulted in trials, the penalties for the perpetrators of these acts were excessively lenient. Despite the interest of Captain-General Francisco Serrano in the first case, Santa Cruz de Oviedo probably got away with only a warning. He was not jailed, and he kept possession of his slaves until the day he died.[66] The mayoral and the ox-driver of the sugar mill San Rafael were sentenced to twelve months in prison and were warned never again to work commanding slaves. They were also ordered to pay the costs of the trial.[67]

These two cases are significant because they took place as slavery was starting to come under fire even within the Spanish empire. During the 1860s, the torture of slaves was regarded as uncivilized and barbarous in cities such as Havana and Madrid. This attitude, however, had not always been so prevalent. The number twenty-five became synonymous with suffering and torture in the first half of the nineteenth century because that was the most lashes a slave could legally be punished with on any given day.[68]

Many slaves did not wait for a miracle to release them from cruelty and servitude: they learned how to denounce their masters, how to claim a new owner or their freedom, and how to use the law in their favor. The case of Florencia Rodríguez, a slave in San Antonio de los Baños, illustrates how slaves began to exploit the colonial law. Florencia's owner, Ramón Sáinz, was renowned for his excesses, his alcoholism, and the sexual favors he obtained from his female slaves. In a letter sent to the captain of the jurisdiction in 1834, Florencia took care to list all of her owner's past abuses. She recalled how Sáinz had had three metallic rings put in the "most secret parts" of her slave friend Inés. She narrated her own sufferings at his hands, including how she had endured "all sorts of things" before he decided to put a silver ring in her private parts as well. Florencia escaped and decided to denounce her master. Her complaints were so solidly argued that even Captain-General Miguel Tacón had to concede her right to look for another owner. Though the records of the trial's outcome are missing, it is clear that Florencia's awareness of the law and her resolution to accuse her owner worked in her favor when the authorities examined the case. Florencia learned how she could use the law, and she did use it effectively to her own benefit.[69]

The cases examined here are only a small sample of the many cases brought to the courts by slaves after excesses were committed against them by their masters and overseers. They are clearly extreme cases. In every one of them, the slaves had no choice but to accuse their masters: their lives were at risk, and an eloquent complaint was probably the only way out for them. Slaves did not always need to be under terrible pressure, however, to turn to available legal channels and cause problems for their masters.

The Counterattack: Slaves and the Law

Although plantation slaves were not often able to take advantage of colonial law in the same way that urban slaves did, they regularly called the attention of the local authorities to their problems. One day in September 1839, a group of thirty-five or thirty-six male slaves from the coffee plantation La Suerte walked to the headquarters of the captain of the jurisdiction of Alacranes, near Matanzas, to complain that they did not have women to share their lives. They stated that since their owner, José de la Luz Piedra, had failed to provide them with that "natural resource," they had decided to present their case to the local authorities. The slaves did not leave the headquarters until they were well satisfied with the promises made by the captain and by their owner to "buy them negras to marry."[70]

Slaves' had numerous reasons to complain to the authorities. In 1828, the marquis of Cárdenas de Monte Hermoso wrote to Captain-General Francisco Dionisio Vives about the "state of insubordination" of forty to fifty slaves from the coffee plantation San José, which was owned by Antonio González Robato. According to the marquis, those slaves had recently tried twice to rebel. Their leader, an African-born slave named Agustín, had gone to the local authorities' headquarters to protest the living conditions on the estate. Soon afterwards, many of the other slaves had followed Agustín to town in order to ask for land to plant their own crops and to demand not to be punished any longer by their mayoral. They threatened to mount an open revolt should the authorities fail to comply with their stipulations. The challenge posed by Agustín and his companions was too serious to be ignored and too bold to be left unpunished. Eleven of the main leaders, including Agustín, were arrested and condemned to different punishments, while the rest of the slaves were returned to the estate.[71]

On exceptional occasions, slaves decided to bypass local authorities and address their objections to higher authorities. Barely a month before the conspiracy of La Escalera was uncovered, and one year after the slave code of Valdés was enacted, thirty-eight slaves from the sugar mill Puerto Escondido walked several kilometers from their sugar mill in the valley of Yumurí to the provincial capital, Matanzas, in order to request a meeting with Governor Antonio García Oña to tell him about the bad treatment they were receiving from their mayoral. Their owner, the count of Fernandina, was called on by the governor to give his opinion in the matter. Predictably, the count supported his employee and blamed the slaves for exaggerating the facts.[72] The mayoral himself stated that the only reason the slaves walked to Matanzas was that he and his men had carried out a raid on the slave quarters, where they had found and confiscated some stolen plantains. The slaves seem to have been tired of the mayoral, and the incident with the plantains triggered their long-lasting discontent, leading them to appeal to the governor of Matanzas. They were returned to the plantation and their complaint was dismissed.[73]

In June 1844, a few months after he had problems with his slaves in Puerto Escondido, the count of Fernandina had to deal with an even more complicated case. A group of his slaves—Fernando Gangá, Justo Gangá, Cayetano Lucumí, Luis Lucumí, Juan Lucumí, and Rita Lucumí—were brought before the Military Commission, accused of offering open resistance to the mayordomo of their tobacco plantation, Santa Mónica. But what initially seemed to be a trial of two Gangá and four Lucumí slaves soon became a criminal process against the mayordomo, Pedro González, and his sons. Curiously, this criminal process was fueled by the count of Fernandina.[74]

In a letter to the authorities dated September 1844, the count himself gave the best account of how events had unraveled.[75] He recalled his visit to the plantation on May 13, during which the mayordomo—who happened to own half of the slaves working on the estate—gave him a detailed report of the situation on the plantation. González then lined up the slaves to meet the count, who wanted to hear their own accounts about the treatment they were receiving from González. Two slaves, Benito and Juan, protested against González's practice of locking them up after sunset in the recently built barracks, and Rita Lucumí complained that while González's recently born slaves received good care, the children of the count's slaves were not properly treated. To prove her point, Rita encouraged the count to ask González why all the children of his slaves were surviving while the count's slave children were dying. After listening to the arguments of Rita and the other slaves, the count asked them to calm down and to obey the orders of his employees. He also promised to have a word with his employees to discuss the situation.

The slaves' complaints were so harsh that the count decided to have a private meeting with González, in which he intended not to ask González about his slave children but rather to prevent the mayordomo from taking any further actions that would generate the slaves' resentment. He reminded the mayordomo that his visit gave the slaves their only chance to voice their distress and improve their lives, and that González therefore should not harbor bad feelings toward them for making complaints. Just when the count thought he had calmed down his slaves, they decided to ask him once again to exempt them from being forced to sleep in the barracks. The count reprimanded them for the second time that day and called upon them to obey their mayordomo without hesitation. That night, the slaves did not show up when called to enter the barracks; instead, they went to sleep in the huts where they had lived before the barracks were built. Once again, the count commanded them to comply with González's orders. He reminded them that González had his personal authorization to punish any slave who offered resistance, and he told them that if they disobeyed again they would be taken to the justice to face trial.

The count finally left later that morning. A few days later, when he was back in his palace in the capital, he received a letter from González informing him that some of his slaves had been killed. Soon, the industrious count had a firsthand account of the events, which he later related in his letter to the authorities:

> That five of González's sons, in his presence, made a horrible butchery with
> my Negroes the day after they were ordered to go to a tobacco house in order

to thread the tobacco, a task for which they needed knives . . . and that after the obedient slaves went to fulfill their obligations, and while they were at work, he [González] and his five sons, carrying their weapons and with their dogs, invaded the tobacco house and, in the middle of such a horrifying assault, summoned the slaves to surrender. The poor slaves, seeing themselves in such an awful situation, ran to the only door of the house, and then the mayordomo's sons murdered three of them [and] seriously wounded another four, who were also bitten and scratched by their dogs, while another nine, probably less frightened, managed to escape by opening holes in the walls of the house, which are made of dried leaves; three or four escaped to give themselves a less miserable death and the rest found shelter in the forests.

The count concluded his account with the statement, "I am persuaded that my Negroes were victims, at the very least, of the barbarous ignorance and perversity of those who were governing them."[76] The six surviving slaves who had been taken to trial were acquitted, and the mayordomo and his five sons were arrested soon afterward.

In this case, the slaves paid a high price for taking advantage of their owner's presence on the plantation to communicate their problems to him. Those who survived were returned to the Santa Mónica tobacco plantation, where they continued to sleep locked in the barracks. Those who were killed and the others who committed suicide paid with their lives for their decision to use a legal—but also perilous—channel in the hope of improving their living conditions. All, however, showed determination and a knowledge of their rights, and they conducted their protests based on this knowledge.

Slaves also addressed authorities individually. The case of José María Lucumí, with which this chapter began, was not unusual. In July 1835, for example, José Antonio Ramos, a slave of Juan Bautista Urrutia, managed to present a request for his right to manumission directly to the captain-general of the island, Miguel Tacón. A few days after Tacón received this demand, Ramos was taken to Pinar del Río, and presumably he was freed soon thereafter in accordance with his right to manumission.[77] Three years later, in June 1838, Francisco Javier Congo tried to claim the same right before the local authorities of Santa Clara. His demand was ignored, and when he was returned to his plantation, he attempted to cut his throat because he was afraid of the punishment he would receive from his master as a reprimand.[78] In another unsuccessful case, Marcial Carabalí fought for his right to be manumitted from 1846 to 1852. His owner was another black man named Gregorio de Zayas. Marcial claimed to have paid four hundred pesos to Gregorio, with which he had bought his manumission. After six years, he lost the case and was returned to his estate in Sitios de Chacón in the locality of El Mariel.[79]

Sometimes, living conditions on the plantations did not allow slaves to learn about and use the law in their favor. In such cases, slaves would resort to the last resort to which they were entitled: the use of padrinos, or protectors. It was a tradition in the Cuban countryside that any slave who committed a crime or made a mistake would find a padrino—preferably but not necessarily a white person—to represent him or her before the owner. Slaves knew about this legal right and frequently took advantage of it. Tomás Ortega told of a representative example in a letter to the governor of Matanzas, Cecilio Ayllón, in 1827. Ortega recalled how one of the slaves on the plantation owned by Gregorio Jiménez asked him to help to get rid of his owner, who often mistreated and beat him. Ortega subsequently asked Jiménez to permit his slave to look for another owner, but Jiménez replied that he should have killed the slave long before and that he would only let the slave go if Ortega agreed to take him to the captain of Lagunillas. This slave, whose name we do not know, managed to escape from his master with the help of a white person who acted as a padrino for him.[80]

On another occasion in 1827, more than thirty slaves ran away from the sugar mill Industria and the coffee plantation Esperanza. The slaves found shelter in the bushes around their estates and waited for their owners to come back so that they could complain about the behavior of their overseers and slave drivers. According to the captain of the jurisdiction, most of them took the precaution of finding padrinos before they consented to give up their resistance and returning to their estates.[81] Similarly, one morning in September 1835 on the coffee plantation San Antonio in Santiago de las Vegas, south of Havana, more than thirty slaves, mostly Africans, escaped to the hills and waited there for their master in order to protest against the "way in which the mayoral was treating them."[82] The slaves came out of the hills only when their owner promised not to punish them for what they had done. They all complained about the bad treatment and poor provisions that the mayoral gave them.[83] The slaves were tried for disobedience and condemned to minor sentences.

What the slaves did in these two cases was to protest—as safely as possible—about their overseers and slave drivers when the occasion seemed propitious. Such behavior was common after the turn of the nineteenth century.[84] When slaves felt that they had the opportunity to bring some distress to the attention of their owners, they often took their chances and went ahead. Testifying within reasonably safe limits provided slaves with an invaluable chance to vent their accumulated anger upon their owners and overseers. Many of the cases I cite in this study serve as clear proof of this opportunism.

Some slaves' desire to bring their oppressors into compliance with the law seems to have been especially strong. In February 1826, for instance, Esteban Mantilla, mayoral of an estate in San Juan y Martínez, killed José de la Trinidad, one of the slaves under his supervision. Contramayoral Antonio Carabalí was the only slave on the estate to back the account offered by the owner, Paulino Iglesias, and the mayoral. According to them, de la Trinidad had fallen, hitting his head on a trunk, and later had died as a result.[85] In contrast, the slave Manuel Iquiabo said that "he had witnessed his [José de la Trinidad's] death, and that the reason was the beating given to him by the mayoral, who hit him four times in his ribs, once in the neck, and once in the head."[86] Antonio Mandinga also confessed to having witnessed the assassination of his comrade.[87] The rest of the slaves—and even other slaves from neighboring plantations—reported that they had seen the mayoral carrying a stick covered with blood and tied to a whip.[88] Because the slaves denounced their mayoral, Mantilla was arrested two months later. Two years later, he was in jail and the case was still under review.[89] A similar case took place some years later in the jurisdiction of San Diego, in which the mayoral, Pedro Luis del Prado, killed a slave using his dogs. Soon afterward, the slaves denounced the mayoral during official interrogations. Since they all declared that they had witnessed the murder, their testimonies were taken seriously and, as in the previous case, the mayoral was arrested and tried.[90]

Conclusion

As this chapter has demonstrated, some slaves in Cuba's western countryside learned of Spanish colonial law (often from their own owners or overseers) and used it in their own favor. Although laws were devised to control them rather than to offer them valid channels to claim their rights, the slaves—

sometimes African-born but more frequently Creoles—made use of every favorable aspect of the law in one way or another. The decision to complain or to claim any right was always a risky one. Slaves who brought their cases before the authorities often ended up frustrated, disappointed, and even punished for their daring actions. It was only after the 1860s—and especially after the passage of emancipation laws—that slaves were consistently able to use the law to their own advantage. But by the late 1840s, slaves knew the law and were able to calculate the risk of their legal actions. Contrary to what has been presumed, they seem to have been keener than ever to claim their rights and to challenge the Spanish colonial legal system. As British consul Joseph Tucker Crawford, an acute observer of Cuban reality, put it in 1848, "They appear to have from day to day less regard for their masters and look for their emancipation most anxiously. This has been very observable since the horrors of the supposed insurrection of 1844, which spread ideas of emancipation all over the island amongst the slaves in a way never entertained by them before."[91]

6

Disguised and Nonviolent
Forms of Resistance

He who has a mouth to tell lies to the other shall lie, and he who has none shall let the world go on.

—Deposition of Felix Carabalí, March 1839

The evening of September 30, 1837, was not an ordinary one for the inhabitants of the small rural village of Güira de Melena. Just after sunset, while some were preparing to go to bed and others were already sleeping, a group of six slaves entered the village, beating African drums and singing songs at full volume. According to a later report by the local captain, Manuel de Jesús Mata, the parade lasted for a considerable time. Mata recalled how the alarmed neighbors "in great numbers woke up and rushed to the streets, some armed with machetes, sabers, and sticks, and others with spears."[1] The twenty-two-year-old leader of the slaves, Anastasio Lucumí, later confessed that without permission they had gathered after 8 P.M. under some palm trees to beat their drums and sing their songs.[2]

In Cuba, as elsewhere, slaves found ways to express their ideas and to reproduce their traditions and inherited knowledge about the world. Even today, African-derived religions and cosmologies constitute an important part of Cuban culture. These cultural traditions and knowledge were transmitted from generation to generation despite the harsh character of the Spanish slave system in the New World. Whether Anastasio Lucumí was aware of it or not, when he and his fellow slaves openly played their music and sang their songs on that evening in 1837, they defied the entire slave system. They ignored the law that required them to apply for official permission before playing music. In an act of even greater daring, they then went public and interrupted the peace of an entire village, causing panic among its inhabitants. Anastasio and his friends shook up the community, alarmed the villagers, and scared the authorities. The authorities retaliated by taking them prisoners and confiscating their drums.

Slaves often re-enacted forbidden habits, customs, religious beliefs, and cultural continuities. Warrior slaves maintained their military pride, for example; not surprisingly, they started numerous movements of resistance. But revolts and marronage were not safe ways to oppose slavery. In contrast, day-to-day life was full of small incidents and events that constituted safer forms of resistance.

Slaves on Cuban plantations were well aware of the limits of their private and public actions. Rather than giving up without offering resistance, they accepted some elements of their oppressors' culture by integrating those elements into their own cultural and religious practices. The best-known example of this phenomenon, though not the only one, is slaves' acceptance of Catholic saints and virgins and their merging of those figures with the various African deities they worshiped. This process, known as syncretism or transculturation, is a fashionable subject of study among scholars today, inspiring prolific research not only in Cuba but also around the world.[3]

Most cases of disguised and nonviolent resistance were probably not recorded by the colonial authorities, but the events that were documented help us to understand slaves' attitudes and their ways of defying the system without getting into too much trouble. Nonviolent or disguised forms of resistance can be seen in a wide range of human behavior, including the sabotage of agricultural tools, arson, feigned sickness, robbery, work stoppages, religious practice, the use of folklore, the spreading of rumors, and many other actions. I do not, however, consider all forms of slave behavior to be acts of resistance. Before concluding that a determined action was an act of resistance, I am careful to ensure that it was intended as such. Though a large number of fires occurred in Cuban sugar mills during the first half of the nineteenth century, for example, I have only analyzed fires that were almost certainly set by slaves.[4]

Sabotage and Other Hidden Forms of Resistance

Slaves practiced various forms of sabotage, which included the destruction of work tools, arson, feigned sickness, dissimulation, false compliance, slander, and foot-dragging. The level of organization and complicity within a group of slaves—established through face-to-face encounters—often determined the degree to which such actions succeeded. Communication between slaves from different plantations was a permanent concern of owners, overseers, and local authorities. In 1843, Carlos Ghersi, a planter and militia officer in the jurisdiction of Macuriges, wrote, "Slaves watch their governors. The houses of the latter are usually far away from the slaves' huts, and thus they make use of the hours of natural rest to run away: they establish communication with other estates, choosing as a meeting point that farm in which the white employees are less vigilant, and from those gatherings and communications are born all the disorders, thefts, and everything else to be feared. . . . They abandon their estates by the footpaths across the hills, cane fields, and coffee dryers, and, therefore, nothing can be avoided."[5]

At such clandestine meetings, slaves chatted, danced, sang, and remembered their beloved African homelands. Some gatherings offered a propitious environment for the development of conspiracies; others ended in huge, almost unstoppable brawls. One such meeting took place during a festivity organized by the slaves of the coffee plantations Nueva Empresa and Asunción in Güira de Melena one night in September 1831. What was meant to be a celebration concluded with a fracas between the slaves of both plantations. In the melee, the two principal combatants and the mayoral of the Nueva Empresa were injured, and the local authorities were forced to intervene in order to restore peace between the slaves.[6]

Planters and local authorities were always keen to keep slaves' clandestine meetings to a minimum. In 1845, Benito García y Santos, a planter in the jurisdiction of Macuriges, wrote to Captain-General O'Donnell, begging him to put an end to "slave liberties." He complained about the lack of barracks at the surrounding sugar mills, where the vast majority of slaves were living in huts. This, according to García y Santos, enabled the free movement of slaves throughout the vicinity and, in consequence, "their participation in meetings, robberies, killings, the concealment of conspiracies, and other disorders of all types and proportions."[7]

There are numerous examples of more refined and disguised forms of communication. Depending on the agreement or laxity of their owners and overseers, slaves might spend time drinking in local taverns or wandering around the vicinity of their plantations selling the products of their conucos.[8] Ignacio Zabaleta Criollo, for example, was punished in April 1844 for selling his plantains without the approval of the mayoral of the sugar mill La Sierra.[9] Clever slaves also used the privacy offered by their huts, and some of them, such as Martín Carabalí of the sugar mill Majana, opened gambling houses and made illegal profits thereby. According to Martín's own testimony, in his hut "people frequently gambled, and there was also a selling point of alcoholic beverages."[10] There is no reason to believe that Martín's hut was an exception; more likely, drinking and gambling were common features of slave life in the nineteenth-century Cuban countryside. In recent archaeological excavations on the coffee plantation Santa Ana de Biajacas, Cuban and U.S. archaeologists have found pipes and "small rounded tokens," which may have been used to play a board game. Huts like Martín Carabalí's became centers where slaves shared their experiences and conspired against their oppressors.[11]

Sabotage was among the most problematic forms of slave resistance for plantation owners, who were well aware of its frequently expensive consequences. Intentional fires were the most feared and costly form of sabotage. Throughout the first half of the nineteenth century, fires were regular occurrences.[12] During uprisings, slaves frequently set the houses and warehouses of their masters ablaze.[13] At other times, they set fire to their owners' property while managing to conceal the source of the blaze.[14]

The testimony given in 1848 by Eduardo Hernández, co-owner of the sugar mill Unión in the jurisdiction of Macuriges, provides a good example of how dangerous and costly fires could be. Although there was no proof that the fire at Unión was set intentionally, the results were devastating. Hernández recalled that "the fire started in the boilers' tower, and from there the flames jumped to the slave huts because the wind from the south was very strong. The sugarcane fields, the slave huts and barracks, and all their pigs were devoured by the flames."[15]

Cases of induced fires were carefully recorded by the colonial authorities, and the majority of them had dreadful consequences for the perpetra-

tors, and often for the owners, who lost crops and buildings in the process. On May 13, 1844, around 11 A.M., two fires started simultaneously in the
sugar chaff warehouse and the wood storage warehouse of the sugar mill
Santísima Trinidad, which was owned by Esteban Santa Cruz de Oviedo.
Barely six months earlier, Oviedo had uncovered an elaborate antislavery
conspiracy at Santísima Trinidad. Even as flames were consuming the warehouses of Oviedo's sugar mill, hundreds of free blacks and slaves were being
interrogated so severely that the conspiracy would later be named for one of
the most notorious instruments of torture used by the colonial officials: the
ladder (La Escalera). Though they doubtless knew the risks they were taking
by setting fires on the plantation, the slaves of Santísima Trinidad chose
to respond to the horrendous treatment they were receiving by destroying
Oviedo's property.[16] All of his detective skills were insufficient to reveal who
among his slaves had destroyed a substantial part of his sugar, chaff, and
wood.[17] Those responsible escaped punishment, while Oviedo was forced
not only to pay for the damages but also to live with the idea that his slaves
were laughing behind his back and, even worse, that they had somehow punished him for his actions.

It has been argued elsewhere that anonymity and mutuality protected
slaves responsible for acts of resistance from punishment. These two things
were the most effective forms of protecting arsonists: if a group of slaves
could set a fire in secret and remained united in refusing to reveal their
responsibility, they would most often escape punishment. In some cases,
however, slave arsonists did not bother to hide their culpability. In 1839, for
example, the slaves from the sugar mill Ingenio Viejo started a massive fire
in their own huts. Francisco Casañas, a neighbor, was absolutely convinced
that the slaves were to blame. The plantation owner, Juan de Dios Casañas,
witnessed them carefully emptying their huts "in anticipation, a fact that
indicates that they were all in concert."[18]

It was also safe—maybe even safer—to act alone, and some fires were
clearly set by individual arsonists. Such was the case with Trinidad, a male
slave of Teresa Hernández, who labored in her sugar mill, Jesús María. In a
letter sent from Sabanilla del Encomendador in February 1832, the captain
of the jurisdiction told the governor of Matanzas about the "adventures" of
Trinidad, who had run away on many occasions. He wrote, "At the moment
I am very nervous, fearing that he [Trinidad] will set fire to the sugar mill . . .

because it would not be the first time he would try."[19] A similar case took place in San Luis de la Ceiba in April 1846 when Pánfilo Carabalí, who had been at large for over nine days, returned to his coffee plantation and, apparently as an act of revenge, burned down his master's house.[20] Pánfilo did not wait to see the result of his daring action: he jumped into a well, where he died. A neighbor ran to the well and spent some time shouting the slave's name, but in response he heard only "the deepest of silences."[21]

In a fairly safe way of resisting their oppression, slaves also employed "slow working" tactics to affect their masters' profits. "Slow working" describes actions intended to delay the processes of production, including foot-dragging, doing things wrong, feigning sickness, and rendering agricultural tools nonfunctional. Scholars of slavery in the New World have often written about slow working, in large part because owners and overseers complained continuously about their slaves' lack of interest in performing their labor. Sometimes owners and overseers even classified the various African ethnic groups according to their supposed predisposition to work. In September 1837, for example, a group of Lucumí slaves rebelled at the sugar mill San Pablo, near the village of Catalina de Güines. After crushing the rebellion, authorities brought some of the rebel slaves to trial. The mayoral of the plantation, Joaquín Curbelo, recalled that he had repeatedly seen José del Carmen Lucumí suggesting to his friends that they disobey the mayoral's orders. Due to José del Carmen's influence, the slaves slowed down their working rhythm, which had a negative effect on the plantation's sugar output.[22]

Slaves knew that foot-dragging and feigned sickness could be highly prejudicial to their masters' pocketbooks. In 1842, Rafael O'Farrill complained about the laziness of his slaves to Captain-General Valdés, writing that "there is not even one slave owner who is not aware that it is better to have two hours of vigorous working than four of laziness."[23] Slaves were so aware of the consequences of their actions that being useless not only became a form of resistance but also a lifestyle of sorts. This was the case for thirty-five-year-old Cristóbal Carabalí, a slave on the coffee plantation Santa Catalina. In 1828, Cristóbal showed no trace of shame when he declared that "he never [did] the work he [was] ordered to do because he [was] lazy." Just before finishing his statement he "confessed, again, that he was lazy and indolent."[24]

Slaves also sometimes ignored orders given by their masters and overseers, even though disobeying orders could lead to tragedy. Despite the advice of his fellow slave Vicente, for example, José del Rosario Criollo decided to ignore an order to feed the horses of his plantation. According to the testimony of Bárbara Gangá, José del Rosario told Vicente that "there will be no problem, that the horses will tell nothing."[25] Ultimately, Francisco, the contramayoral, showed up and asked José del Rosario why the horses had not been fed. The Creole slave replied in an "improper" manner, and a fight began. The results were appalling. José del Rosario killed the contramayoral and injured his wife, Guadalupe Mandinga. Months later, he was sentenced to death by hanging; after he was dead, his body was mutilated, and his right had was exhibited in "the most public place."[26]

Cuban slave owners often associated slow working with the relaxation of plantations' internal control and discipline. In 1843, for example, the owner of the sugar mills La Arratía and Achuri, Salvador de la Paz Martiartu, was sued by his neighboring planters for being too relaxed with his slaves. The slaves of the Achuri were renowned for defying orders and even for daring "to kill some pigs of their owner in front of his own eyes."[27] Pedro Cano, one of the litigating planters, testified that Achuri's 179 slaves were all well-dressed and fed and that "when the mayorales whipped any slave, they were fired by the owner without any further examination."[28]

Coffee plantations were frequently singled out as having a more relaxed slave-labor system than the harsh dawn-till-dusk work that characterized the sugar plantations. Based on his own experience, the influential planter Wenceslao de Villaurrutia wrote in 1842 that "most of the rebellions that occur in our fields originate from good treatment, rather than by rigor."[29] He continued:

This is the reason why those movements have taken place more often in cafetales, where the work is much less active than in the ingenios. The vigilance over the slaves and the exigencies of owners and mayorales are also less, and the slaves are generally better fed and rested. [These revolts] have usually started during holidays, when the slaves had the opportunity to dance and to celebrate, in the wrong assumption that these concessions would make them happier. A notable example of this type took place in the Sumidero, where Mr. Fouquier and his family were all assassinated by his own negros, among whom he lived as a kind father rather than as a master. He used to persuade his slaves to do their work rather than to force them to, and he also used to stimulate them with rewards, promising manumission or freedom in his testament to those who behaved better during their time as his slaves.[30]

Not surprisingly, slaves may have taken advantage of the relaxation of work discipline on coffee plantations to strengthen their social networks and also to conspire and reproduce their hidden transcripts.[31]

Disobedience also became a recurrent nonviolent form of collective resistance against rules that slaves considered unfair. Throughout the first half of the nineteenth century, slaves resorted to peaceful work strikes with increasing frequency. These strikes were invariably provoked by changes in the estate's government that altered the slaves' way of life or their working schedules. Collective protests against overseers were a common feature of day-to-day life on nineteenth-century Cuban plantations.[32] On August 12, 1843, the seventy slaves of the coffee plantation Asunción, which was owned by the marquis of Aguas Claras, refused to work and asked their owner for a new mayoral. After calling for the local militia, the marquis ordered twenty-five lashes to be given to each of the slaves before locking them up in the plantation's barracks.[33]

The introduction of a new overseer sparked protest on numerous occasions. In 1828, the slaves of the coffee plantation Santa Catalina, near Guanajay, decided to strike against the newly appointed mayoral in order to protect their private space and some established privileges. According to the slaves, the main reason for their protest was the refusal of the new mayoral, Antonio Toscano, to provide them with "a proper lunch." Other important objections were that he "was not letting them smoke their pipes and chat, and that when they were returning from carrying out their physical needs he often whipped them."[34] Something similar happened on the coffee plantation Jesús Nazareno, near Bejucal. A few days after the owner, María de la Luz Valdés, appointed a new mayoral, the slaves became openly insubordinate, declining to work and challenging the authority of the mayoral, who apparently was totally inexperienced in the government of slaves and, as he himself recognized, was "nothing but a poor man."[35]

Sometimes, slaves did not wait to find out how good or bad a new mayoral would be before refusing to obey his orders. At the sugar mill Nueva Vizcaya in the jurisdiction of Yumurí in 1837, the owner, Santiago Garmony, lined up the slaves to introduce them to their new mayoral. Questioning the wisdom of Garmony's choice, the slaves straightforwardly rejected the new man's authority. According to the mayoral's later testimony, "Once the slaves gathered together to recognize him, one of them—small and fat—stepped out of the line and told him that they did not want him because they knew that he was a wicked man; and immediately afterward, some other slaves exclaimed at once that they did not want him. Seeing their resolution not to have a mayoral or anyone else to govern them but their owner, [the mayoral] was asked by the latter to go away until further notice."[36]

In all these cases of peaceful protest, the slaves succeeded in achieving their aims, and their opinions were heard and considered. But when collective protests did not work out, the results were often appalling. At the sugar mill Santísima Trinidad in 1834, a multiethnic protest threatened not only the stability of the estate but also the peace of the entire region. This protest was sparked not by a new mayoral but by a change in work practices. In his desire to have a good crop, the mayoral decided to wake up the slaves earlier in the mornings and force them to work on Sundays. When the slaves complained to the mayoral about the new schedule, they were seized and punished, and one of them, twenty-six-year-old Lucas Congo, died as a consequence of the injuries he received while resisting the mayoral and his men.[37]

Though this strike failed, most slave strikes stood a good chance of succeeding, making them a relatively safe way for slaves to contest authority and to achieve change. Although in some cases they escalated due to the deeply rooted oppressive character of the Spanish slave system in Cuba,[38] peaceful insubordination within the confines of the plantation was always controlled to a certain degree by the internal rules of each estate. It was when discontent was translated into nighttime raids and road assaults that slave resistance became harder for the authorities to deal with. In these circumstances, planters, plantation employees, and their white neighbors became almost powerless. Apart from building slave barracks, there was little they could do to prevent such attacks.

Cuban archives contain also several cases of cattle theft recorded by the Spanish authorities.[39] In 1836, this accusation was leveled against Antonio Viví of the sugar mill San Miguel and his comrades Elías Gangá and Bernabé Congo from the adjacent sugar mill Mogote in Guanabo. Although the slaves denied all the charges, evidence pointed to their culpability in the theft and killing of an ox belonging to the San Miguel.[40] In another case, José Manresa, who lived in the village of Aguacate, saw a slave riding a mare in October 1836. When he asked the rider to identify himself, the slave abandoned the mare and disappeared behind a stone wall. Manresa seized the mare and continued alongside the road when, suddenly, he was attacked by two slaves who "threw stones at him in such a manner that he was forced to flee."[41] According to Manresa's testimony, one of the attacking slaves was the same slave who had previously been riding the mare.

The slaves Manuel Congo and Simón Arará were sentenced in 1837 to wear irons for two years for assaulting Francisco Díaz and his slave Ignacio Congo with a machete on the road to Guanajay. Although their owner wrote a letter to the president of the Military Commission in Havana defending his slaves, they were found guilty and were severely punished.[42] A similar event took place in Santa María del Rosario one night in 1833, when Manuel Cabrera, a soldier attached to the garrison of Güines, was attacked by three slaves. Cabrera was traveling in the company of free black Francisco Estrada, who was also assaulted. During the subsequent interrogations, he recalled how he and Estrada were stopped by three "negroes" who told them, "Hey you, pieces of dog, stop there and get yourselves facing down." Then they were deprived of all their belongings, including the official correspondence that Cabrera was taking to Güines. The investigation was thorough, and the three muggers were accused and convicted later that year.[43]

While many assaults and robberies did not result in fatalities, there were cases in which robbery led to killing and killing led to capital punishment.[44] In November 1845, four slaves residing on different estates in the jurisdiction of La Guanábana attacked a paddock located on land of the sugar mill Santa Lutgarda. Two slaves guarding the entrance of the paddock were murdered: Roque Congo was hanged, and Miguel Gangá was lethally wounded with a machete or a knife. The thieves took "three dozen hens."[45] Shortly before, they had also stolen two mares from a neighboring estate in order to facilitate their assault and escape. José Benigno Criollo and José María Criollo, the ringleaders, were sentenced to death. The other two slaves, Casimiro Criollo and José de la Cruz Congo, were sentenced to eight and six years' imprisonment respectively.[46] In a similar case, Claudio Carabalí, a slave of the sugar mill Jesús María in Guanabacoa, was attacked by three slaves from a neighboring estate when he found them stealing a bag of bananas from his sugar mill's provisions.[47] Claudio received multiple machete injuries and died days later of tetanus. The three thieves were believed to belong to the nearby sugar mill La Chumba. They were never identified or brought to justice for their crime.[48]

Robberies and road assaults carried out by slaves were such common events that authorities took steps to prevent them.[49] The need for such official steps was supported by the opinions of foreign visitors, who frequently made references to the dangers of traveling on Cuba's roads.[50] Owners, overseers, and local authorities echoed these views.

In exceptional cases, the advantages offered by darkness were used for robberies of another type. The theft in 1844 of a cage containing the head of a mulatto executed for his involvement in the conspiracy of La Escalera is a good example of how unpredictable robberies could be.[51] And, as we shall see, nighttime was also the perfect setting for cultural forms of resistance.

Culture and Resistance on Cuban Plantations

Slaves had many other ways of expressing discontent and offering resistance. To confront their oppressors within safe limits, they reproduced those aspects of their traditional cultures that were not totally forbidden by their masters. On many occasions, singing and dancing or even attending mass were acts of resistance circumspectly performed before the eyes of their masters and overseers.[52]

Conversations—often in African languages—were undoubtedly the most frequent and potentially dangerous form of everyday resistance among slaves. While not all conversations were part of the day-to-day resistance practiced by the slaves, many of them were certainly intended to protest, criticize, or convene forms of dissent. These conversations often took place in safe spaces where the collective hidden transcript was enacted, such as the abovementioned hut of Martín Carabalí. From the mid-eighteenth century, slaves heavily outnumbered white masters and employees in the Cuban countryside.[53] This allowed them to have moments of privacy in which they could speak freely, voicing their ideas and complaining about their immediate problems to their fellow slaves, often in African languages unknown to their overseers and owners. Sugarcane and coffee fields were propitious places to spread rumors and to gossip about maroon leaders like Yará Gangá, Domingo Macuá, or José Dolores Congo. One such maroon hero, José Ramón Mandinga, played an important role in the revolt of Guamacaro in June 1825. While most of the rebels were captured or assassinated, the one-handed José Ramón escaped to the nearby mountain range of El Sumidero, where he remained at large for over three years. During this time, his fame grew among the slaves on plantations in the vicinity. When he was finally captured in 1828, public prosecutor Francisco Seidel referred to his "legend" in the following terms: "Taking into consideration his reputation among the slaves of being almost impossible to catch due to his speed and intelligence in the mountains; also considering that since he took part in the uprising two years ago he has been at large despite our efforts to capture him; and due to his constant communications with the slaves, whom he repeatedly tried to entice from their masters, he must be beheaded, his right hand cut off, and both shall be exhibited in the most public place in the roads of the mountain range of El Sumidero, until time will consume them."[54]

From the mid-1820s, slave owners began to construct barracks—barracones—to keep slaves locked up at night in order to avoid nocturnal meetings between slaves from different estates.[55] These barracks produced an unwanted secondary effect, however: confined within these spaces, slaves were able to foster kinship ties and to create a relatively private economy beyond the surveillance of their overseers. Some of the major slave revolts of the period were conceived and planned within slave barracks. Most important, slaves had the freedom to gossip and complain within the safety of their confinement.[56]

In their legal testimony, slaves frequently referred to their conversations in the barracks, fields, and processing installations of the plantations. In 1844, Matías Gangá, a slave of the sugar mill Santa Ana de Jaspe, told prosecutor Felipe Arango that "when they were grinding the sugar cane, he and his comrades used to complain a lot in their own language about the punishments they were frequently receiving."[57] It was no secret to masters and overseers that slaves had private spaces where they talked and gossiped, often under the influence of tobacco and alcohol or while playing games and gambling. Authorities and planters frequently made reference to slaves' use of private spaces and to the danger of letting slaves chat and gossip behind their backs. They knew how dangerous simple talk and seemingly harmless rumor could become. Consequently, they never stopped worrying and trying to control slave life to avoid potentially fatal consequences.

In 1842, the wealthy planter Joaquín Muñoz Izaguirre reminded Captain-General Valdés that the slaves "know about everything, that they talk, make their comments and often speak out double-meaning words."[58] Another Creole planter, José Manuel Carrillo, backed Izaguirre's opinion. He commented that "it is compulsory not to forget the suspicion of the slaves, always ready to interpret in their own favor whatever measure can offer them some independence and to undermine the authority of the owners and employees on their estates."[59]

José Leopoldo Yarini, an Italian physician and planter who suffered alongside his slaves during the deadly cholera epidemic of 1833, commented on the daily behavior of the slaves of his sugar mill in Guamacaro. About communications—specifically communications during the epidemic— Yarini wrote: "The useful, hardworking, and intelligent Negroes watch with an increased attention the dangers that threaten them. As a result, they acquire some ability to evaluate their circumstances with some precision and order; they compare and draw natural conclusions. . . . They talk about the dangers and get news from outside, which happens to be always exaggerated because of the reigning fear."[60] As Yarini suggested, the rumors that spread among rural slaves were likely to be exaggerated, and consequently they could provoke excitement among the slaves. The correspondence and testimony of authorities and plantation owners clearly reveals that rumors were a constant source of concern to them.

There is strong evidence that slaves plotted secretly in their own languages prior to the revolts of Guamacaro in 1825 and Guanajay in 1833. In the first case, many slaves declared that the three leaders of the rebellion used to meet on Sundays to discuss private matters in their African languages, keeping their conversations secret from the other slaves. Several slaves made similar statements after the revolt of El Salvador was put down by the authorities. It seems to have been the case that the old Lucumí leaders on the plantation plotted together with the leader of the recently arrived Lucumís to "murder all the whites and return to their homeland."[61]

Conversations were often accompanied by physical gestures. In the stressful environment of the plantations, even grimacing could be interpreted as an infraction serious enough to justify punishment. At least in one case, it led to the murder of a slave. In 1845, the mayoral of the estate La Llanada in Pinar del Río, José Ignacio Padilla, was charged for his excesses in punishing the slave Alejo. In his defense, Padilla argued that Alejo was not carrying out the task he was ordered to do. Padilla recalled how he tried to compel the slave to work, only for Alejo to "threaten him with treacherous and insubordinate glances."[62] Padilla's response was to crack the whip, and a fight began in which Padilla ended up injuring the slave twice with his machete.[63] In analyzing the responses of slaves to slavery in a later period (1860–1886), Rebecca Scott cited a letter written by the British consul in Sagua la Grande, in which he commented on how the local planters were complaining about the fact that their former slaves were laughing "in their faces."[64] Slaves could resist their oppression, then, with their angry or grinning faces.[65] Sometimes they could also do so by speaking out in apparent innocence, as we will see later in this chapter.

It might seem unusual to suggest that music was a form of resistance. Nonetheless, it is impossible to deny that on some occasions singing, dancing, and playing the drums helped slaves to bear the heavy weight of their condition, to keep alive their African cultures, to express discontent, and to offer a counter-argument to slavery. Music was an essential part of African slaves' lives. They sang, danced, and played their drums when they were happy, when they were sad, when they were angry, when they went to war, and even when they were desperate.[66] In consequence, although not all incidents of slave singing, dancing, and drumming were acts of resistance, it is highly likely that many of the slaves' song lyrics, their gestures of dance, and both the sacred and profane language of their drums were conceived and performed with the intention of resisting the rules to which all slaves were subjected.[67] Singing, dancing, and drumming were common to some of the most important and largest slave revolts of the period. They were present in the revolt of Guamacaro in June 1825, and they played a very important role in the uprising on the coffee plantation Salvador in 1833. In the latter case, some of the lyrics sang by the rebels were transcribed by colonial prosecutors, one of the few examples of surviving African lyrics known to be transcribed in Cuba in the first half of the nineteenth century.[68] Slave drumming parties—batucadas—also constituted a source of concern to Brazilian authorities, who sometimes saw them as "a breeding ground for rebellion and slave autonomy."[69]

In a documented case from 1839, nine Lucumí slaves from the sugar mill La Conchita were repressed—some of them were killed—after enjoying what the estates's employees called "an unusual happiness."[70] Cleto Lucumí, one of the surviving slaves, later indicated that their mood was far from happy. On the contrary, he explained, they were singing a very sad song: "[We] do not see father anymore, [We] do not see mother anymore."[71] In other words, they were singing, dancing, and swinging their machetes in the presence of the mayoral at the same time that they were complaining in their own language about the sorrows slavery had caused them.

The sound of African drums—which were often associated with slave uprisings—frequently frightened authorities and slave owners. Drums had a significant role in communication between estates, and to play them was considered a privilege among slaves. In 1844, Manuel Lucumí was supposed to play the drum during a planned rebellion that was uncovered by the colonial authorities. According to some of his comrades, he was expected to do so because he was "a drum player ever since they were making war in Africa."[72]

Drumming sessions could last for hours and sometimes even for days. This seems to have been the case in the jurisdiction of Juan Angola, Matanzas, in 1827. In May, the captain of the jurisdiction had to answer a letter from the central government in Havana that asked for an explanation of the drumming parties that were taking place in Juan Angola, some of which were said to last for over thirty-six hours, attracting large numbers of slaves.[73] Eleven years later, Captain-General Joaquín de Ezpeleta had to deal with this issue once again. He received a letter detailing the anxiety that drumming parties were causing whites on neighboring plantations in the Cuban countryside. The author, a planter named José María Gavilán, complained that he was "convinced about the threat posed by the drumming parties for big and small countryside estates, because when the scandalous noise begins, slaves travel from the farthest of distances, and thus the parties are followed by robberies, drunkenness, and other subsequent problems."[74]

Very much in the same manner, sayings—refranes—could be a means of resisting oppression from a more or less safe position. Slave sayings have been passed down by oral tradition and have been recorded by scholars such as Fernando Ortiz, Lydia Cabrera, and Samuel Feijóo.[75] African slaves brought some of these sayings with them from Africa and created many others while living on the plantations. These sayings addressed almost every aspect of their lives as slaves, from religion and punishment to revolt and marronage.

In certain circumstances, sayings constituted the safest method for slaves to discuss the hardships of their day-to-day life without risking their lives in the process. Sayings such as "when the owner holds the whip is better to smile at him" and "mayoral cracks the whip and everybody goes quiet" were crystal-clear ways of expressing fear and caution. These sayings were sometimes complemented by other, bolder ones, such as, "a machete can also cut the master" and "for more peaceful than a slave can be, if the whip hurts him, he will shake it off." Runaway slaves and their experiences were part of this peculiar body of knowledge. Most of these sayings were not meant to spur open resistance, but rather to discourage slaves from facing the deprivations of life as a maroon: "with good words you get the maroon out of the woods" and "where there is a dog, there are not maroons."

Sayings reaffirming the slaves' dignity abounded. Some of them were related to their warrior heritage, such as "the sons of Oggún do not run away from the battle," or the Carabalí saying, "only that one who has no heart does not go to war." Others were of a more philosophical nature: "a good heart does not understand of [skin] color," and "the rich man can buy the slave but not his soul."

Overall, slave sayings were part of slaves' day-to-day life. They were spelled out in almost every occasion and within every environment. They were a good way to complain about slavery and on occasion to make light of problems, a behavior that has become an intrinsic part of Cuban culture with the passing of time. In fact, many of these sayings can be heard even today in Cuba. Some of them are quite popular: "there is neither a brave black nor a sweet tamarind," for example, or the classic "the totí is always to blame for everything."[76]

Slave gatherings provided a propitious environment for the reproduction of African or transculturated religious beliefs, otherwise known as "black witchcraft." In January 1845, some slaves and a white neighbor of San Diego de los Baños were brought before a justice, accused of practicing black magic. According to the owner of the estate where the incriminating ceremony took place, the white neighbor was acting as the chief of the after-midnight gathering in a coffee field, while three slaves were burning gunpowder, talking in foreign languages, and reciting "sweet potato field, that we will not lack sweet potatoes; plantain field, that we will not lack plantains."[77] Don Antonio Alvarez, the owner of the estate, testified that the slaves "were making weird gestures" and that the entire ritual seemed to have been a "witchcraft affair."[78]

Religious practices and beliefs were deeply embedded in the daily life of sugar and coffee plantations and, presumably for that reason, they were not habitually recorded. Only on extraordinary occasions did captains of a jurisdiction begin criminal proceedings against sorcerers and practitioners. A truly unusual case occurred in 1839 outside the village of Aguacate, located between Matanzas and Madruga. On March 11, the free pardo Cirilo César walked to the lieutenancy of Aguacate at 6 A.M. to give notice about some things he had seen at a crossroad while on his way to work that morning. An official visit to the crossroad by Lieutenant Luis de León confirmed what César had told him, and de León took the unusual step of writing a detailed description of his findings. He and his men discovered a "dead pigeon flanked by some silver coins, negroes' kinky hairs, and some pigeon feathers."[79] Soon thereafter, they discovered a rustic litter made of wild cane—caña brava—and a piece of brand-new fabric on which a dead, colorful rooster was lying. The bird had been wrapped in an old shirt; his chest had been cut open, and inside there were more silver coins. Another search in the nearby bushes revealed another litter and a knife without a haft that was jammed into the ground and surrounded by some silver coins. Before leaving the crossroad, de León ordered his men to check the area one last time. This time, they found a large number of black beans that formed a trail which crossed the village and disappeared on its other side.[80]

There were more surprises in store for Lieutenant de León. At noon, María de los Angeles Alfonso came to him carrying a brand-new pan made of clay with a dead white pigeon inside. The white feathers of the pigeon's tail had been replaced by black feathers taken from another pigeon. The pan had two small spots of blood inside it, along with some splashes of whitewash.[81] These items had all been found at the same crossroad by her son earlier in the morning. At this point, de León feared that some sort of slave movement was underway. His investigations quickly led him to the coffee plantation Concepción, owned by Lorenzo Xiqués and renowned across the region for its lack of discipline. The slaves, most of them Carabalí Isuamos, were allowed to visit the village on Sundays to do their own shopping. The results of de León's inquiry led to the arrest of some of the Concepción slaves. Surprisingly, they did not attempt to hide the truth, but rather cooperated with de León and the prosecutor, Apolinar de la Gala. José Carabalí Isuamo was the first to declare that his companions—Cleto, Esteban, and the contramayoral Félix, all Carabalí Isuamos—were responsible for the artifacts found at the crossroad. He mentioned their frequent gatherings and their private talks in the Carabalí language. He recalled seeing the others going to Aguacate with the intention of buying pigeons, as well as dividing the money that they would use later in the "witchcraft." José also said that he had attempted to discourage them, warning that "those were not good things." Félix's reply embodied the very concept of disguised resistance. He warned José: "He who has a mouth to tell lies to the other shall lie, and he who has not any shall let the world go on."[82] Félix's concise response was about lies, about keeping one's mouth shut, but above all it was about protecting the sacred space that offered slaves the opportunity to resist the oppression to which they were subjected in myriad ways. In simple terms, Félix was offering two options to his fellow slave: he could either lie or refrain from saying anything, but he should never denounce his comrades. If he was not brave enough to participate in the others' activities, he should remain silent and "let the world go on."

José's testimony also revealed interesting details about the African meaning of the discovered items. He said that in Africa,

> they used to put the head of a dead cock in those places where there were people buried; that the knife nailed to the earth is usually placed on top of the graves of those who die with the reputation of being brave men, in order to stop them from coming out of the ground to take the others with them. That in his land, when they want to fight, they put in an old basket piquant leaves, ants, heads of nasty dogs, snakes, and all the warm things they can find, and that they take them to the place where they are going to fight as a signal of challenge, and also as a signal that those who are going to fight are brave men.[83]

After taking the testimony of several men and consulting higher authorities, prosecutor Gala decided to charge the three main protagonists, including the contramayoral, with practicing witchcraft. He asked for sentences that ranged between four and six months in irons for Cleto, Esteban, and Félix, "due to the moral influence that these Negroes, called sorcerers by the other Africans, exercise[d] over the spirits of the most stupid, and because [Spanish] laws recommend the correction of this sort of occurrence."[84]

Death had an extremely important meaning for all African slaves. Respect for the dead and belief in reincarnation were widespread among the many ethnic groups that arrived in Cuba. Not surprisingly, then, slaves rejected orders that involved handling the dead and sometimes clashed with owners and overseers over issues related to the dead. At the sugar mill Santa Ana de Jaspe in March 1844, for example, the slaves strongly resisted burning the grave of a slave girl who had died earlier that year.[85]

Perhaps the best first-hand account of the attitudes toward the dead that Africans brought with them to Cuban plantations was given by the planter José Leopoldo Yarini. The cholera epidemic of 1833 offered Yarini an exceptional opportunity to take a close, almost anthropological look at the lives of his African slaves. He took note of almost every aspect of their existence, and the dead became a central issue in his writing. Among his day-to-day observations, he recorded a burial ceremony: "They roughly construct a box and then put many plantain leaves inside; they then place the hat on the head of the dead man and wrap him with a blanket. They also place inside [the box] a bottle of sugarcane brandy, a pipe with tobacco . . . a club formerly owned by the mayoral, and many cock feathers [laid] over his face and chest. . . . The most extravagant thing I saw was that they put on his chest a cross made by holy palm leaves and various images of the Virgin and the saints."[86]

After witnessing this intriguing ceremony, the Italian doctor and planter did his best to find out the meaning of what he had seen. He asked many of his slaves—almost certainly Carabalís—about it before one of them finally satisfied his curiosity. That slave explained:

> In a remote province in the depths of Africa and only for those who die in a distant place, they celebrate such a ritual. When the dead man returns to life in his own country, he needs the club to defend himself from the dogs, because they might not recognize him. The sugarcane brandy and the tobacco are for the journey. The cocks' feathers mean that at dawn he must be back in his land in order to wake up his friends and relatives for them to recognize him and admit him again into the family after his captivity. Then he begins a new epoch of life, and he enjoys again peace and domestic union with his gods.

Yarini continued his account: "The same Negro who explained to me those African mysteries experienced a sweet satisfaction while he talked about them. I noticed happiness in his face and sparkles in his eyes while he related that he will be reborn and that he will enjoy his freedom once again."[87]

To prevent death, African slaves resorted to all kinds of magic artifices. Amulets were especially valued for their ability to preserve life and to make men invisible to their enemies.[88] Yarini observed the use of amulets among his slaves and wrote about their belief in the supernatural powers of these talismans:

> Some of them carried necklaces, the so-called brujerías, amulets they believe [make them] infallible to survive the cholera. Others carried another sort of neck-

laces, like rosaries made of holy palm leaves, with many little crosses of wood. Later I found out that they were made of *guásimas*, because this is the favorite tree for those who decide to kill themselves by hanging. Many wore pieces of glass, birds' feathers of different colors, snails, dogs' fangs, cocks' spurs, and dried weeds with Negroes' kinky hairs and seeds of Palma Christi. . . . A Negro of the Gangá nation was carrying a little *majá* or *jubo* [snake] wrapped in a handkerchief and covered by a cap. . . . When I saw him he gave me the cap with extreme care and recommended that I not kill that animal, because it was the safest remedy against the evil herbs that other Negroes use to kill people.[89]

The many forms of resistance practiced by the slaves on Cuban plantations constituted an essential part of their everyday lives. By engaging in these practices, slaves managed not only to reproduce their own cultures in their new "homes," but also to give substance to their resistance to slavery.

Conclusion

There are undoubtedly countless aspects of slave life still waiting to be un-covered. There are no known accurate descriptions of specific religious cer-emonies, for example, and much remains unclear about the ways in which slaves translated their hidden transcripts into public transcripts. There are still many gaps in our understanding and knowledge about interethnic rela-tions on plantations, and major uncertainties remain even about the place of origin of some ethnic groups. Specific issues—such as the reproduction of gossip and rumors and their transformation into myths, the meanings of life and death, and especially the importance of kingship to African-born slaves and their descendants—call for further study to enable us to reach a better understanding of slaves' daily life and the forms of resistance they practiced.

While there is little doubt that robbery, collective protest, sabotage, and other hidden violent actions were forms of resistance, many scholars still question the categorization of other hidden forms of resistance, which did not involve violent practices against the dominant authorities. Were these actions truly ways of resisting domination?

At this point, it is useful to remember the words of James C. Scott: "A con-vincing performance may require both the suppression or control of feelings that would spoil the performance and the simulation of emotions that are necessary to the performance."[90] Slaves accepted—or were forced to accept—their oppressors' saints and virgins. They often attended—or were forced to

attend—Catholic mass. No one questions, however, that a process of transculturation took place in various slave societies across the New World: slaves
enacted a convincing performance of worshiping San Francisco de Assis,
Santa Bárbara, and the Virgen de la Merced when in fact they were adoring
Orula, Shangó, or Obatalá, regardless of what they called them. It remains
unclear whether colonial authorities, slave owners, and overseers consciously
contributed to the continuous reproduction of African beliefs or believed
that their mission to Christianize the slaves was a successful enterprise.

Similar arguments might be made with respect to singing, drumming,
and dancing. Apart from a few exceptional cases, plantation owners and employees barely knew any words of the African languages their slaves spoke.
How, then, could they know what the slaves were singing in their presence?[91]
Even more difficult to interpret were the contortions of their bodies and the
secret and profane meanings of the sound of their millenarian drums. In
the slaves' private spaces, jokes, gossip, rumors, complaints, and euphemisms
were frequently—but not always—conscious forms of resistance that also
constituted an implicit rejection of slavery.[92]

Conclusion

Slavery distorts the personality and all human relationships, so that only through resistance
can the self be realized and dignity restored.

—Michael Craton, "Proto Peasant Revolts"

The attempts made by Cuban ruling groups from the end of the eighteenth
century to replace neighboring Saint Domingue as a producer of sugar and
coffee had long-lasting consequences for Cuba and its inhabitants. The elite's
need for a cheap labor force prompted them to bring large numbers of African slaves to the island. The increasing slave population, in turn, became
a serious threat to the prosperity of these same men and their descendants.
From the late eighteenth century, they lived in a constant state of fear. As
Arango y Parreño cried before the Spanish courts in 1811, the matter of slave
imports deserved all their attention, since it affected their lives, their fortunes, and the lives and fortunes of their descendants.[1]

Although slave revolts and marronage were the most frequently mentioned and feared forms of resistance, they were not the sole concerns of Cuban slave owners. Authorities, planters, and the white population in general
also saw danger in the French and Haitian revolutions, in the French invasion of the Spanish peninsula under the command of Napoleon Bonaparte,
and particularly in the British abolitionist movement. All these things, they
believed, threatened their prosperity and wealth. Nonetheless, slave unrest

was by far the most serious issue they had to deal with. Whites observed how slaves learned of the existence of laws that they could use to their own advantage—laws that entitled them to manumission and coartación. They also noticed that slaves, both African-born and Creole, did not happily accept their captivity in the way their owners would have liked. In a sense, slave owners became captives of the slave system they had created.

Throughout most of the twentieth century, scholars—in particular Cubans— paid more attention to violent forms of slave resistance— namely, rebellions and marronage—than they did to day-to-day forms of resistance. However, as Carolyn E. Fick pointed out some years ago, "If one attempts to quantitatively register the extent of resistance in any given slave society by merely counting the number of openly organized revolts, or the number or duration of large maroon communities such as the quilombos, then one risks looking in the wrong direction."[2] In this study, following Fick's advice, I have done my best to look beyond revolts and marronage. Slaves resisted their bondage in a wide variety of ways, which were determined by the living conditions on their plantations and by their individual backgrounds. In Cuba, as elsewhere in the Americas, the environment of oppression on a given plantation often determined the extent of slave resistance. Harsh rules could lead to slave revolts, but they could also prevent their occurrence; yet, according to various prominent slave owners, relaxed organization could have the same results. It was a common belief among planters and overseers that the relaxation of control on coffee plantations was the reason that most of the slave uprisings of the first half of the nineteenth century took place there rather than at sugar mills, where control and working conditions were more strict.[3]

Eventually, slaves developed alternative forms of resistance within the boundaries of the Spanish colonial system. The cases presented in Chapters 5 and 6 of this work suggest the diverse range of actions undertaken by slaves to improve their existence or to retaliate against different types of oppression. They claimed some fundamental rights and pushed the limits of the legal system in a genuine process of negotiation. To some extent, this negotiation might be understood as accommodation. In order to make substantial changes in their lives, slaves who were not ready to rise against their owners and overseers instead played along with the rules of the insular slave system. The day-to-day social interactions among slaves, overseers, owners, and authorities fostered important relationships that offered slaves the opportunity to instigate qualitative changes without losing their lives in the process. Eugene Genovese stated some years ago that "accommodation itself breathed a critical spirit and disguised subversive actions, and often embraced its apparent opposite—resistance."[4] Understood in this way, legal actions taken by slaves—as well as their many disguised cultural and non-violent acts—might be classified as incidents of both accommodation and resistance. Since these actions contributed "to the cohesion and strength of a social class threatened by disintegration and demoralization," however, and since they frequently bordered on being violent actions, they are all considered as forms of resistance in this study.[5]

Chapters 5 and 6 showed us how nonviolent and disguised forms of resistance were part of quotidian life on western Cuban plantations throughout the first half of the nineteenth century. The same can be argued for other places in the New World. As João José Reis and Eduardo Silva have pointed out in reference to Brazil, the vast majority of the slaves "either by strategy, creativity, or luck were living in the best possible ways."[6] In his study of slave resistance in Puerto Rico, Luis Figueroa has similarly argued that "slaves also deployed adaptations that did not necessarily seek the short-term destruction of slavery's social relations but provided time and space for autonomous existence and helped reaffirm their humanity in the face of the normative rules of chattel property imposed by slavery."[7]

Throughout this work it has been stressed that this "adaptive resistance" was also present in the daily life of African and Creole slaves on Cuban plantations. That is not to say, however, that Cuban slaves did not take advantage of the system when possible, very often with the clear intention of undermining its basis and transforming it.

In the safety of kitchens, huts, and barracks, and alongside roads, slaves discussed topics that affected their lives. They never abandoned their religious beliefs, not even when they merged their own gods with Catholic virgins and saints. They held on to their mother tongues despite adopting foreign words and accents with the passage of time. They never forgot their own cultures' methods of socialization, their musical instruments and traditional dances. Eventually, in autonomous events—private talks, religious ceremonies, and festivities—they created safe spaces where they voiced their concerns and, more important, organized common strategies in their struggle against oppression. In the safety of spaces like these, the slaves of the coffee plantation Salvador conceived the rebellion that caused panic among the population of the villages of Banes and Guanajay and the surrounding area in August 1833. In spaces like these, maroons were welcomed back to plantations, and it was there that their stories of freedom in the forests were told and repeated among the slaves. It was in spaces like these, away from their masters and overseers, that slaves reproduced their hidden transcripts. It is clear that the reproduction of the slaves' hidden transcripts was instrumental for the development of a "distinctive subculture characterized by a strong 'us against them' social imagery."[8]

When we address the meanings of negotiation or accommodation, particularly for slave societies in the Americas, we cannot presume that these categories have definitive meanings. Slaves' everyday actions—including some forms of negotiation and reluctant acquiescence—in no sense demonstrated their "acceptance of slavery."[9] Gossip that was spread to undermine the authority of overseers or slave drivers, jokes that aimed to make a mockery of masters or their families, and complaints directed at provoking more complaints were all day-to-day forms of disguised resistance in the Cuban countryside.

Scholars of slavery in the Americas, depending on the quality of their sources, lately have tried to address these non-politically driven forms of resistance. Many, however, still refuse to give credit to any form of adaptive resistance and continue to underrate it as a phenomenon. African slaves did not always have in their minds risking their lives and the lives of their relatives in what many arguably considered futile attempts to fight their social condition in a straightforward manner. Instead, they learned to survive. And in the process, they learned how to preserve their dignity and to resist oppression in safer ways.

Chapters 2, 3, and 4 addressed slaves' violent and open forms of resistance. Along with rebellions and escapes, I have included suicides and homicides in this study. The evidence offered in Chapter 4 suggests how widespread and serious a problem slave suicides were for Cuban slave owners by the mid-1840s. It also reveals that acts of collective suicide were performed on various occasions by African-born slaves. Although slaves' reasons for committing suicide might not be clear, the historical sources allow us to know what the public opinion of the time had to say about these deaths, as well as slaves' beliefs in the afterlife.

Chapter 2 considered the phenomenon of the homicides of masters, overseers, and other white people in the western Cuban countryside. The act of taking someone else's life—especially when that person was in charge of restricting slaves' movements and granting them concessions—seems to have been common on nineteenth-century western Cuban plantations. Although there are no statistics available to clarify the extent of these actions, many cases were registered in one way or another by travelers, authorities, and newspapers. I have focused principally on rebellions in that chapter, however, arguing that African-led revolts preponderated over other types of movements, particularly conspiracies and revolts led by Creoles and free colored people. The extraordinary sequence of slave movements that took place on the plantations of western Cuba between 1795 and 1844 is comparable only to that which occurred in Salvador de Bahia and its surroundings between 1807 and 1835.[10] The vast majority of these movements were led by African-born men who fought their battles according to their African knowledge of war. With a few exceptions, most of these movements had an intrinsically African character and were not the result of events that occurred elsewhere. As was shown in Chapter 1, African-born slaves had enough knowledge of warfare to fight their own battles without needing any French revolutionary, Haitian ex-slave, or British abolitionist to teach them that slavery was not a good thing and that it was simple logic to offer resistance to it, both as a social system and as a personal tragedy.

In Chapter 3, I have suggested that marronage has been overemphasized in Cuban historiography. In Cuba, as elsewhere in the Americas, slaves escaped to the mountains, forests, and swamps searching for freedom—or at least for autonomy. Runaway slaves and maroon communities, although they never constituted a serious threat to the status quo of the island, were often a matter of discussion in the meetings of the colonial authorities in Havana and Santiago de Cuba; they were also a source of irritation for planters and the rural population in general. Although they never threatened the colonial establishment as they did in places like Jamaica, Demerara, and northeastern Brazil, their continuous presence constituted a problem that never really disappeared in the first half of the century. Although they were a minority, "their significance far transcended their numbers."[11]

Slaves' experiences were dissimilar and particular. Every plantation had specific rules, spatial distributions, and working schedules that could and did affect slaves' lives and methods of resistance. The behavior of masters, overseers, and slave drivers, as well as the sociocultural background of slaves, determined to a considerable extent the forms resistance took. Born on the island, Creole slaves learned Spanish as children. Some were taken away from their mothers and raised by slave nannies chosen by owners and overseers. In this practice can be seen elements of the slave owners' belief that slave children should "never be under the direction of their parents," as the care of slave children was "a matter for the owners."[12] Creoles' experiences, then, were intrinsically Cuban and did not necessarily involve traumatic incidents other than those attached to their subordinated social condition. Their understanding of Spanish helped them better to perceive what was happening around them. They knew and understood the concerns of their oppressors and often were familiar with the things that provoked their masters. Many of them knew about the Great Slave Revolution in Saint Domingue and about British attempts to bring the slave trade to an end. More important, Creole slaves were more likely than African-born slaves to comprehend and make use of Spanish colonial law. As in other New World slave societies, such as Brazil, most of the Cuban court cases in which slaves denounced their masters and claimed or asked for the right of manumission or coartación were brought by Creole slaves.[13]

African-born slaves, in contrast, all went through the traumatic experience of the Middle Passage. They were all survivors, the fittest among the many men and women sold on the shores of the distant African continent. Many of them had been slaves even before leaving Africa. Many brought their military experience with them to the New World, as well as their beliefs in reincarnation—a continuous source of trouble for slave owners and authorities in Cuba, who never quite managed to understand what this "Pythagorean resurrection" was about. African-born slaves were forced to settle down, to learn a new language, and to answer to new names. The slave life that was imposed on them was not easy to accept. It is not surprising, then, that they were more prone to resort to violent forms of resistance than Creole slaves. The sequence of slave revolts that occurred in Cuba between 1795 and 1844 had an intrinsically African character, although Creole slaves and freemen were frequently involved in these revolts as well. As Chapter 2 has demonstrated, most of these slave movements were organized and led by African-born slaves, who often had arrived on the island only recently. Domingo

Aldama, one of the most conservative planters of mid-nineteenth century Cuba, put it simply when he said that African-born slaves "have very different ideas and customs to those of the white race . . . and those habits and facts that seem very normal to the nature of white men could be a torment for African Negroes . . . because they are not in harmony with their customs, with the memories of their childhood and their native land; memories that are enjoyable for the freemen, and consequently must be much more pleasing for the slaves."[14]

Creole slaves were slaves from the moment they were born, but for thousands of African-born men and women freedom was an unforgettable experience they were keen to regain and for which they were willing to die. This is not to say that Creole slaves were ignorant of the meaning of freedom or that they were not disposed to fight to change the course of their lives, as in fact they often did. But for many Creole slaves, the question was not if they would be set free, but when and how. Once slaves began to make use of the law, it was only a matter of time before popular knowledge of how to obtain a life of freedom was fairly widely disseminated. Manumission and coartación were viable legal channels for Creole and urban slaves, but for most African-born slaves—particularly those who worked long hours on remote sugar plantations—they were almost nonexistent possibilities. African-born slaves had no one to negotiate with and did not know what to negotiate in any case. Revolts, marronage, and suicide became their most effective forms of resisting their oppressors.

Being a Creole or an African was a crucial factor in determining the path of resistance that slaves were likely to take. Life in cities, towns, small holdings, and plantations was predictably different and, to a significant extent, could affect the choices made by slaves. On plantations, other important factors weighed on the slave minds, including the rigor of the surveillance to which they were subjected, the ethnic composition of the slave population within their plantations and in their vicinity, and the geographical location of the holdings.

Assessing the various forms of resistance practiced by slaves on western Cuban plantations is a demanding enterprise. I have tried to show, sometimes through the very voices of the slaves, that accommodation or adaptation did not necessarily meant acceptance, but rather were tame yet conspicuous ways of resisting domination. Although from the mid-1840s onwards very few slave rebellions were recorded and marronage became an increasingly rare phenomenon, other forms of slave resistance lived on. Slave suicides, for example, continue to be a concern for slave owners and authorities throughout the 1850s and 1860s, right until the end of the Ten Years' War and the final abolition of slavery in 1886. Adaptive resistance flourished precisely after the mid-1840s, when what Manuel Moreno Fraginals often referred as the period of "good-treatment" began.[15] Slaves never stopped resisting slavery as a condition or as a system; the evidence is clear in the historical sources for anyone interested in finding it.

Although I have been able to look at hundreds—maybe thousands—of cases of resistance, I am aware of the shortcomings of the available documentation and of the many historical gaps that call for different approaches. I wish I could have offered greater insight into the role played by slave women, for example, and also by the elderly. Sources suggest that their participation in plots, marronage, and other acts of resistance was far more important than we might think today. Unfortunately, source materials, which often focus on adult male slaves, do not offer adequate information about the involvement of women or the elderly in the various forms of resistance described in this study.

My biggest frustration has been having no choice but to work with the Christian names given to slaves in the New World, rather than with their original names. The name issue is particularly important when assessing the lives of African slaves, because in West and West-Central Africa names have meanings that refer to local, personal, or familial characteristics. Knowing African names thus would have helped me to understand the social place of those slaves that we know today only as Félix Carabalí, Pablo Gangá, or Lorenzo Lucumí. In only a few cases, Spanish authorities recorded the original names of slaves. Perhaps future research will help us to understand better who these men and women were, where exactly they came from, what their beliefs were, and what they knew or believed about the world. For the moment, at least, we have the opportunity to examine some of the actions by which these slaves were immortalized—actions that on many occasions cost them their lives.

In these pages, I have tried to offer a fresh look at the forms of resistance practiced by slaves across western Cuba in the first half of the nineteenth century. By analyzing not only those conventionally accepted and studied forms of resistance (such as marronage and rebellions) but also nonviolent and disguised forms of resistance, I have also attempted to offer a more comprehensive picture of the slave universe—a picture full of contradictions, limitations, and liberties that we do not comprehend very well even today. I have been keen to accentuate the differences between African-born and Creole slaves because I believe these differences constituted one of the main factors in determining the forms in which slaves reacted against slavery. African-born slaves who arrived in western Cuba managed to keep their languages and cultures alive, although they suffered changes during the long-lasting process of transculturation. This does not mean, however, that slaves changed because they changed the places where they lived. No matter how many stories and descriptions they heard, Creole slaves could only imagine what their African ancestors' homes looked like. They were the ultimate products of an extraordinarily bitter experience, but they were also the seeds of a new epoch. African-born slaves, in contrast, especially those who experienced the Middle Passage as adults, remained Africans until the day they died. Neither the transatlantic voyage nor the experience of being forced to resettle far from the world they had known could make them into different people. The distinctions between African-born and Creole slaves need to be clearly assessed in future studies. We must make a greater effort to comprehend their different mentalities and cultures and to remember always that slaves were real people, not fictional characters.

Finally, I feel obliged to state that this work, beyond its intrinsically academic objective, is meant to be a sort of modest homage to the African-born and Creole slaves who, through their actions, made history some two hundred years ago. I am proud to have been involved, even at a considerable remove, with these men and women and to have served as a link between them and the present. I hope this study not only helps us to understand past events more clearly but also illustrates the endurance of the human will even under the harshest conditions of existence.

Glossary

ALMAGRE Red ochre powder.

AYUNTAMIENTO Town hall or city council. It was used to designate both the location of council meetings and the group of people who comprised the council.

BARRACÓN Barracks used to accommodate slaves on sugar and coffee plantations throughout Cuba from the first half of the nineteenth century.

BATEY Central living space on sugar and coffee plantations. The batey was the place where slave houses were usually located; it was also the assembly point for the slaves before and after going to the fields.

BOZAL Recently landed or not yet seasoned slave.

BRUJERÍA Any non-Catholic religious act carried out by the slaves.

CAFETAL Coffee plantation.

CAÑA BRAVA *Bambusa vulgaris*. Large tree well-known across Cuba. It is commonly found on the banks of rivers, creeks, and lagoons.

CIMARRÓN Runaway slave.

COARTACIÓN Process by which slaves were able to buy their freedom, paying predetermined sums of money over a period of time.

COARTADO   Slave who benefited from coartación.

CONTRAMAYORAL   Slave driver. Slave drivers were often chosen from among a plantation's slaves; they could also be free people of color or whites.

CONUCO   Garden plot granted to the slaves by their owners, where they could grow products to sell.

GUÁSIMA   Guazuma ulmifolia lam or Guazuma tormentosa. The wood of this tree or shrub was used to make various artifacts, including chairs and shoe spurs. Its fruits were often used to feed pigs and cattle.

GÜIRA   Crescentia cujete. This common tree was grown in Cuba because of its fruits, which were used to make maracas, cups, jars, and vases.

INGENIO   Sugar mill.

JUBO CUBANO   Alsophis cantherigerus. Cuban snake, which can reach up to nearly five feet in length. It mostly lives along the coasts of Cuba but can be also found in inland areas. Its saliva is slightly toxic.

MACHETE   Cutlass.

MAJÁ   Cuban snake. The most common type is the famous majá de Santa María, or Cuban boa (Epicrates angulifer). It can reach up to thirteen feet in length and can be found in holes and rock piles all over the island.

MANSO   Slave who did not escape or rebel.

MAYORAL   Plantation overseer.

MAYORDOMO   Plantation manager.

PADRINO   1. Godfather; 2. Free person—usually white—chosen by the slaves to intercede with their owners on their behalf.

PALENQUE   Runaway slaves' villages in the mountains, swamps, or forests. Similar to Brazilian quilombos.

PALO DE CUAVA   Also known as cuaba de ingenio or white ironwood (Hypelate trifoliata). It is a species of tree belonging to the soapberry family found in the southern part of North America and in the Caribbean. It can grow up to thirty or forty feet tall. In Cuba it is still used today as a ritual plant in Yoruba ceremonies.

PARDO Offspring of a white and a mulatto couple.

RANCHEADOR Slave hunter.

RANCHERÍA Runaway slaves' village, smaller and less protected than a palenque.

SÍNDICO PROCURADOR Colonial official in charge of interceding on behalf of the slaves.

TRAPICHE Sugar-grinding millhouse.

VEGA Tobacco plantation.

Notes

INTRODUCTION

1. Eugene Genovese, *From Rebellion to Revolution: Afro-American Slave Revolts in the Making of the Modern World* (Baton Rouge: Louisiana State University Press, 1979), 3, 36–38, 82–85, 91–92.

2. As Gloria García has shown in her groundbreaking study of the slavery in Cuba, from the 1850s onwards Creole slaves began to appear more often in the primary sources claim-

ing their rights, a fact that reflects the increase of the Creole slave population after the 1827 census. Gloria García, *La esclavitud desde la esclavitud: La visión de los siervos* (Havana: Ciencias Sociales, 2003).

3. John Thornton, *Africa and the Africans in the Making of the Atlantic World, 1400–1800* (Cambridge: Cambridge University Press, 1992), 320.

4. See J. R. Montalvo, C. de la Torre, and L. Montané, *El cráneo de Maceo. Estudio antropológico* (Havana: Imprenta Militar, 1900); Fernando Ortiz, *Los negros brujos* (Madrid: Libreria de Fernando Fe, 1906); Ulrich Bonnell Phillips, *Life and Labor in the Old South* (Boston: Little, Brown, 1930); Gilberto Freyre, *The Masters and the Slaves [Casa-Grande & Senzala]: A Study in the Development of Brazilian Civilization* (Berkeley: University of California Press, 1986). Some more recent studies on the topic include: Vera Kutzinski, *Sugar's Secrets: Race and the Erotics of Cuban Nationalism* (Charlottesville: University Press of Virginia, 1993); Aline Helg, *Our Rightful Share: The Afro-Cuban Struggle for Equality, 1886–1912* (Chapel Hill: University of North Carolina Press, 1995); Alejandra Bronfman, "Reading Maceo's Skull (Or the Paradoxes of Race in Cuba)," Princeton University, Program in Latin American Studies Boletin (Fall 1998); Alejandro de la Fuente, *A Nation for All: Race, Inequality, and Politics in Twentieth-Century Cuba* (Chapel Hill: University of North Carolina Press, 2001).

5. Examples of this racial conflict include the War of the Independent Party of Persons of Color in Cuba in 1912 and the frequent violent racial conflicts in the southern U.S. states, such as Rosewood and the 1917 hangings in St. Louis.

6. Raymond A. Bauer and Alice H. Bauer, "Day-to-Day Resistance to Slavery," *Journal of Negro History* 27 (1942): 388–409.

7. Herbert Aptheker, *American Negro Slave Revolts* (New York: Columbia University Press, 1943).

8. José Luciano Franco, *La conspiración de Aponte* (Havana: Publicaciones del Archivo Nacional de Cuba, 1963); C. L. R. James, *The Black Jacobins: Toussaint L'Ouverture and the San Domingo Revolution* (New York: Vintage, 1963).

9. Monica Schuler, "Day-to-Day Resistance to Slavery in the Caribbean during the Eighteenth Century," *Bulletin of African Studies Association of the West Indies* 6 (1973): 57–75; Stuart Schwartz, "Resistance and Accommodation in Eighteenth-Century Brazil: The Slaves' View of Slavery," *Hispanic American Historical Review* 57 (1977): 69–81.

10. Genovese, *From Rebellion to Revolution*, 2–3.

11. This debate has continued in recent years. See, for example, David Geggus, "Slavery, War, and Revolution in the Greater Caribbean, 1789–1815," in *A Turbulent Time: The French Revolution and the Greater Caribbean*, ed. David Barry Gaspar and David P. Geggus (Bloomington: Indiana University Press, 1997), 1–50; Matt D. Childs, *The 1812 Aponte Rebellion in Cuba and the Struggle against Atlantic Slavery* (Chapel Hill: University of North Carolina Press, 2006), 155–88; Manuel Barcia, "Slave Rebellions in Latin America during the 'Age of Revolution': Bahia and Havana-Matanzas from a Comparative Perspective" (M.A. thesis, University of Essex, 2002), 5–24, 64–83.

12. For an insightful analysis of the difference between the British West Indies and the rest of the Americas, and particularly Brazil and Cuba, see Michael Craton, *Testing the Chains: Resistance to Slavery in the British West Indies* (Ithaca: Cornell University Press, 1982).

13. See, for example, John K. Thornton, "African Dimensions of the Stono Rebellion," *American Historical Review* 96:4 (1991): 1101–15; Thornton, "African Soldiers in the Haitian Revolution," *Journal of Caribbean History* 25:1 & 2 (1991): 58–80; Thornton, *Africa and Africans in the Making of the Atlantic World*; Paul Lovejoy, *Transformations in Slavery: A History of Slavery in Africa* (Cambridge: Cambridge University Press, 1983); Lovejoy, "The African Diaspora: Revisionist Interpretations of Ethnicity, Culture, and Religion under Slavery," *Studies in the World History of Slavery, Abolition, and Emancipation* 2:1 (1997). Online at: www2.h-net.msu. edu/~slavery/essays/esy9701love.html. The most recent work on this topic is Verene Shepherd and Glen Richards, eds., *Questioning Creole: Creolisation Discourses in Caribbean Culture* (Kingston: Ian Randle, 2002).

14. Robert Paquette, *Sugar Is Made with Blood: The Conspiracy of La Escalera and the Conflict between Empires over Slavery in Cuba* (Middletown, Conn.: Wesleyan University Press, 1987); João José Reis, *Slave Rebellion in Brazil: The Muslim Slave Uprising of 1835 in Bahia* (Baltimore: Johns Hopkins University Press, 1993). See also Gabino La Rosa Corzo, *Runaway Slave Settlements in Cuba: Resistance and Repression*, trans. Mary Todd (Chapel Hill: University of North Carolina Press, 2003).

15. Craton, *Testing the Chains*; David Barry Gaspar, *Bondmen and Rebels: A Study of Master-Slave Relations in Antigua* (Baltimore: Johns Hopkins University Press, 1985).

16. James C. Scott, *Weapons of the Weak: Everyday Forms of Peasant Resistance* (New Haven: Yale University Press, 1985); Scott, *Domination and the Arts of Resistance: Hidden Transcripts* (New Haven: Yale University Press, 1990). According to Scott, subordinates are not passive spectators of the events that happened in their lives. Rather, they develop different forms of response to the relations of power in which they are living. In order to understand and explain these practices, Scott argues that there exists a public transcript, intrinsically related to the public realm and found under the virtual control of the dominant groups, and a hidden transcript, followed by subordinate groups (as well as dominant ones), which was located within the safe limits of their respective private spheres. Scott, *Domination and the Arts of Resistance*, 4–5.

17. Emilia Viotti da Costa, *Crowns of Glory, Tears of Blood: The Demerara Slave Rebellion of 1823* (Oxford: Oxford University Press, 1994); Robert Paquette, "Social History Update: Slave Resistance and Social History," *Journal of Social History* 24:3 (1991): 681–85.

18. Some of these works, among many others, are: Roger N. Rasnake, *Domination and Cultural Resistance: Authority and Power among an Andean People* (Durham, N.C.: Duke University Press, 1988); Richard G. Fox and Orin Starn, eds., *Between Resistance and Revolution: Cultural Politics and Social Protest* (New Brunswick, N.J.: Rutgers University Press, 1997); Steven Gregory, *Santería in New York City: A Study in Cultural Resistance* (New York: Routledge, 1999); Daniel E. Walker, *"No More, No More": Slavery and Cultural Resistance in Havana and New Orleans* (Minneapolis: University of Minnesota Press, 2004).

19. This has also been the case in scholarship on other slave societies in the New World. Stuart Schwartz, for example, has recently noted that in Brazil the study of slave resistance has been dominated by the subjects of quilombos and revolts. Schwartz, "Cantos and Quilombos: A Hausa Rebellion in Bahia, 1814," in *Slaves, Subjects, and Subversives: Blacks in Colonial*

Latin America, ed. Jane Landers and Barry Robinson (Albuquerque: University of New Mexico Press, 2006), 247.

20. See, for instance, Francisco González del Valle, La conspiración de La Escalera (Havana: El Siglo XX, 1925); Elías Entralgo, "Los problemas de la esclavitud: la conspiración de Aponte," Cuadernos de Historia Habanera 12 (1937).

21. See, for example, M. L. Pratt, Imperial Eyes: Travel Writing and Transculturation (London: Routledge, 1992). The term and its author have been also included in a compilation of postcolonial concepts. See Bill Ashcroft, Gareth Griffiths, and Helen Tiffin, Key Concepts in Post-Colonial Studies (London: Routledge, 1998), 233–34.

22. See Pedro Deschamps Chapeaux, El negro en la economía habanera del siglo XIX (Havana: UNEAC, 1971); Deschamps Chapeaux, Los batallones de pardos y morenos libres (Havana: Instituto Cubano del Libro, 1976); Lydia Cabrera, Cuentos negros de Cuba (Havana: Ediciones Nuevo Mundo, 1961); Cabrera, El Monte: Igbo, finda, ewe orisha, vititinfinda (Miami: Rema Press, 1968); Cabrera, Koeko Iyawo: Aprende Novicia. Pequeño tratado de regla Lucumi (Miami: Ultra Graphics, 1980); Cabrera, Vocabulario Congo: El bantú que se habla en Cuba (Miami: Peninsular Publishing, 1984).

23. Raúl Cepero Bonilla, Azúcar y abolición (Havana: Ciencias Sociales, 1961); Manuel Moreno Fraginals, El ingenio: Complejo económico social cubano del azúcar, 3 vols. (Havana: Ciencias Sociales, 1978).

24. There are countless examples of this tendency. See, for example, Miguel Barnet, ed., The Autobiography of a Runaway Slave: Esteban Montejo, trans. Jocast Innes (New York: Pantheon, 1968); Jorge Ibarra, Ideología mambisa (Havana: Instituto del Libro, 1967); Luis Estévez y Romero, Desde el Zanjón hasta Baire: Datos para la historia política de Cuba (Havana: Imprenta La Propaganda Literaria, 1899).

25. Ortiz's last work was Historia de una pelea cubana contra los demonios (Havana: Ucar, García, 1959), a study about a controversial episode of demonic possession in the seventeenth-century Cuban countryside.

26. See José Luciano Franco, Esclavitud, comercio, y tráfico negreros (Havana: Archivo Nacional de Cuba, 1974); Franco, Contrabando y trata negrera en el Caribe (Havana: Ciencias Sociales, 1976); also the classic of Juan Pérez de la Riva, El barracón y otros ensayos (Havana: Ciencias Sociales, 1975).

27. João José Reis and Eduardo Silva, Negociação e conflito, 32 (my translation).

28. See Gloria García, La esclavitud; Rafael de Bivar Marquese, Idéias sobre a administração das plantations escravistas nas Américas, séc. XVII–XIX, Relatorio de Pesquisa no. 4 (Sao Paulo: FAPESP, 2000); Manuel Barcia, Con el látigo de la ira: Legislación, represión, y control en las plantaciones cubanas, 1790–1870 (Havana: Ciencias Sociales, 2000). See also Alan Watson, Slave Law in the Americas (Athens: University of Georgia Press, 1990); Paul Finkleman, ed., Slavery and the Law (Madison: University of Wisconsin Press, 1997); Manuel Lucena Salmoral, Los códigos negros de la américa española, 1768–1842 (Alcalá de Henares: UNESCO/Universidad de Alcalá de Henares, 2003).

29. These works have mainly studied the last period of the slave regime, a time of change in the practices of domination over slaves. See María del Carmen Barcia, Burguesía esclavista y abolición (Havana: Ciencias Sociales, 1987); Rebecca J. Scott, Slave Emancipation: The Transition to Free Labor, 1860–1899 (Princeton: Princeton University Press, 1985); Marquese, Idéias.

30. See Paquette, *Sugar Is Made with Blood*; Childs, *The 1812 Aponte Rebellion*; David Geggus, "Slave Resistance in the Spanish Caribbean in the Mid-1790s," in *A Turbulent Time*, ed. Geggus and Gaspar, 131–55.

31. See note 2.

32. The most recent work on day-to-day slave life in eighteenth- and nineteenth-century Cuba is María del Carmen Barcia, *La otra familia: Parientes, redes, y descendencia de los esclavos en Cuba* (Havana: Casa de las Américas, 2003).

33. Klein, *Slavery in the Americas*; Gwendolyn Midlo Hall, *Social Control in Slave Plantation Societies: A Comparison of St. Domingue and Cuba* (Baltimore: Johns Hopkins University Press, 1971).

34. Franklin Knight, *Slave Society in Cuba during the Nineteenth Century* (Madison: University of Wisconsin Press, 1970); Verena Martínez Alier, *Marriage, Class, and Colour in Nineteenth-Century Cuba: A Study of Racial Attitudes and Sexual Values in a Slave Society* (Cambridge: Cambridge University Press, 1974); Scott, *Slave Emancipation in Cuba*; Paquette, *Sugar Is Made with Blood*; Allan J. Kuethe, *Cuba, 1753–1815: Crown, Military, and Society* (Knoxville: University of Tennessee Press, 1986). Since the early 1990s, some other studies have addressed this issue, notably Philip Howard, *Changing History: Afro-Cuban Cabildos and Societies of Color in the Nineteenth Century* (Baton Rouge: Louisiana State University Press, 1998), and Maria Elena Diaz, *The Virgin, the King, and the Royal Slaves of El Cobre: Negotiating Freedom in Colonial Cuba, 1670–1780* (Stanford: Stanford University Press, 2002).

35. Among others, categories of dividing, understanding, and defining the content of human resistance have been "passive-active," "hidden-public," and "overt-covert."

36. Although conspiracies are intrinsically secret, nonviolent forms of resistance, their ultimate aim is a violent one. This is why I have included them among forms of violent resistance. I have proceeded in a similar way when examining petit (mostly nonviolent) and grand (mostly violent) marronage.

37. In addition to the British consuls and their teams, the British were represented in Cuba by royal officials of the Courts of Mixed Commission for the Abolishment of the Slave Trade (from 1817) and by several merchants and ship captains who left accounts of their travels to the island.

38. Winthrop D. Jordan, *Tumult and Silence at Second Creek: An Inquiry into a Civil War Slave Conspiracy* (Baton Rouge: Louisiana State University Press, 1993), 95.

39. This is a constant throughout the period. The vast majority of the testimonies obtained after African-led revolts were consistently unanimous about the motives that led the rebel slaves to rise against their owners or overseers.

40. Conspiracies such as the Aponte rebellion in 1812 and La Escalera in 1844 are transparent examples of the extent to which reasons for violent resistance could vary and depended on the characteristics and backgrounds of the leaders of slave rebellions in Cuba during the first half of the nineteenth century.

41. The same problem can be found in accounts produced by contemporary eyewitnesses, whose personal opinions were more often than not biased against slaves.

42. See the deposition of Fernando Mandinga. Revolt on the cafetal Empresa, 1840. ANC: GSC. 939/33131.

43. Costa, *Crowns of Glory, Tears of Blood*, 74.

44. Even the harshest critic of the agency of subordinate groups, Gayatri C. Spivak, does not deny that subalterns can speak and that indeed they do speak, even when their words are expressed through the language or voice of the dominant group. See Spivak, "Can the Subaltern Speak?" in *Marxism and the Interpretation of Culture*, ed. Cary Nelson and Lawrence Grossberg (Urbana: University of Illinois Press, 1988), 271–313; Rosalind O'Hanlon, "Recovering the Subject: Subaltern Studies and Histories of Resistance in Colonial South Asia," *Modern Asian Studies* 22:1 (1988): 189–224.

45. Carlo Ginzburg and Carlo Poni, "The Name of the Game: Unequal Exchange and the Historiographic Marketplace," in *Microhistory and the Lost Peoples of Europe: Selections from Quaderni Storici*, ed. Edward Muir and Guido Ruggiero (Baltimore: Johns Hopkins University Press, 1991), 8.

46. Ginzburg and Poni have rightly noted that "criminal and inquisitional proceedings are the closest thing we historians have to the modern anthropologist's field study." Carlo Ginzburg and Carlo Poni, "Il nome e il come: scambio ineguale e mercato storiografico," *Quaderni Storici* 40 (1979): 181–90.

47. Edward Muir, "Introduction: Observing Trifles," in *Microhistory and the Lost Peoples of Europe*, ed. Muir and Ruggiero, xii.

48. Carlo Ginzburg, "Microhistory: Two or Three Things That I Know about It," *Critical Inquiry* 20 (1993): 28. In this article (24), Ginzburg reaffirms his stance on this issue: "The hypotheses, the doubts, the uncertainties become part of the narration."

CHAPTER 1

1. José Luciano Franco, *Comercio clandestino de esclavos* (Havana: Ciencias Sociales, 1980); David Murray, *Odious Commerce: Britain, Spain, and the Abolition of the Cuban Slave Trade* (Cambridge: Cambridge University Press, 1980); Laird T. Bergad, Fe Iglesias García, and María del Carmen Barcia, *The Cuban Slave Market, 1790–1880* (Cambridge: Cambridge University Press, 1995).

2. See Franco, *Comercio clandestino de esclavos*; Murray, *Odious Commerce*.

3. Louis A. Perez, *To Die in Cuba: Suicide and Society* (Chapel Hill: University of North Carolina Press, 2005), 35.

4. Gwendolyn Midlo Hall, *Slavery and African Ethnicities in the Americas: Restoring the Links* (Chapel Hill: University of North Carolina Press, 2005), 68.

5. Robin Law, *The Oyo Empire, c. 1600–c. 1836: A West African Imperialism in the Era of the Atlantic Slave Trade* (Oxford: Oxford University Press, 1977), 23. See also Law's most recent article on this topic, "Ethnicity and the Slave Trade: 'Lucumi' and 'Nago' as Ethnonyms in West Africa," *History in Africa* 24 (1997): 205–19.

6. Hall, *Slavery and African Ethnicities*, 22–54.

7. See Michael A. Gomez, *Exchanging Our Country Marks: The Transformation of African Identities in the Colonial and Antebellum South* (Chapel Hill: University of North Carolina Press, 1998), 173–74; Hall, *Slavery and African Ethnicities*, 22–54; Matt D. Childs, "The Defects of Being

a Black Creole: The Degrees of African Identity in the Cuban Cabildos de Nacion, 1790–1820," in *Slaves, Subjects, and Subversives*, ed. Landers and Robinson, 209–46.

8. Paul E. Lovejoy, *Slavery, Commerce, and Production in the Sokoto Caliphate of West Africa* (Trenton, N.J.: Africa World Press, 2005), 55.

9. See Chapter 6. Case of the rebellion of 1833 on the cafetal El Salvador in Guanajay. I would like to thank Olaniyi Rasheed and Elisée Soumonni for helping me understand the meaning of many of the names of the rebel slaves of the cafetal El Salvador.

10. See the case of Bozen Mandinga Moro. ANC: ME. 713/B.

11. For Lucumi-Nagô-Yoruba slaves in the New World, see Chapter 2. Also see Howard M. Prince, *Slave Rebellion in Bahia, 1807–1835* (Ph.D. diss., Columbia University, 1972); Reis, *Slave Rebellion in Brazil*; Maureen Warner Lewis, *Guinea's Other Suns: The African Dynamic in Trinidad Culture* (Dover, Mass.: Majority Press, 1991); Lewis, *Trinidad Yoruba: From Mother Tongue to Memory* (Tuscaloosa: University of Alabama Press, 1996); Toyin Falola and Matt D. Childs, eds., *Yoruba Diaspora in the Atlantic World* (Bloomington: Indiana University Press, 2005).

12. Cited by Robin Law, "A West African Cavalry State: The Kingdom of Oyo," *Journal of African History* 16:1 (1975): 7, 14. For a comprehensive study of the horse in West Africa, see Law, *The Horse in West African History* (Oxford: Oxford University Press, 1980).

13. See Robin Law, "The Constitutional Troubles of Oyo in the Eighteenth Century," *Journal of African History* 12:1 (1971): 25–44.

14. Ibid., 26–27.

15. See, for example, Stuart B. Schwartz, *Sugar Plantations in the Formation of the Brazilian Society: Bahia, 1550–1835* (Cambridge: Cambridge University Press, 1985), 475; Reis, *Slave Rebellion in Brazil*, 93–96.

16. H. U. Beier, "Spirit Children among the Yoruba," *African Affairs* 53:213 (1954): 329.

17. Ibid. See also Philip Koslow, *Yorubaland: The Flowering of Genius* (New York: Chelsea House, 1996), 20.

18. P. S. O. Aremu, "Between Myth and Reality: Yoruba Egungun Costumes as Commemorative Clothes," *Journal of Black Studies* 22:1 (1991): 8.

19. It is very likely that Aja, Yoruba, and Ibo slaves were also traded to the New World under the term "Arará."

20. See Harold Courlander, "Gods of the Haitian Mountains," *Journal of Negro History* 29:3 (1944): 339–72; Maya Deren, *The Voodoo Gods* (London: Paladin, 1975); Walter Mosley, "Voodoo," *Callaloo* 38 (1989): 153–55; Janice Body, "Spirit Possession Revisited: Beyond Instrumentality," *Annual Review of Anthropology* 23 (1994): 407–34; Suzanne Preston Blier, "The Path of the Leopard: Motherhood and Majesty in Early Danhome," *Journal of African History* 36:3 (1995): 391–417; Bob Corbett, "Introduction to Voodoo in Haiti," http://www.webster.edu/~corbetre/haiti/voodoo/overview.htm.

21. Elisée Soumonni, *Daomé e o mundo atlântico* (Amsterdam: SEPHIS, 2001), 50–60.

22. Ibid. See also Phillip D. Morgan, "The Cultural Implications of the Atlantic Slave Trade: African Regional Origins, American Destinations, and New World Developments," *Slavery & Abolition* 18:1 (1997): 129.

23. See, for example, Carolyn E. Fick, *The Making of Haiti: The Saint Domingue Revolution from Below* (Knoxville: University of Tennessee Press, 1989); Fick, "The Saint Domingue Slave Insurrection of 1791: A Socio-Political and Cultural Analysis," *Journal of Caribbean History* 25:1 & 2 (1991): 1–40; John K. Thornton, "African Soldiers in the Haitian Revolution," *Journal of Caribbean History* 25:1 & 2 (1991): 58–81; David Geggus, "Slavery, War, and Revolution in the Greater Caribbean, 1789–1815," in *A Turbulent Time*, ed. Gaspar and Geggus, 1–51. In the New World, Arará slaves were always related to obscure magical rituals, and especially to the practice of Voodoo. The fact that human sacrifices were common events in Dahomey, perhaps like no other West African kingdom, also helped to boost the myth of the Ararás. See James, *Black Jacobins*; Guérin Montilus, *Dieux en diaspora: les loa haïtiens et les vaudou du royaume d'Allada (Bénin)* (Niamey: CELHTO, 1988); John K. Thornton, "On the Trail of Voodoo: African Christianity in Africa and the Americas," *The Americas* 44:1 (1988): 261–78; David Geggus, "The Bois Caïman Ceremony," *Journal of Caribbean History* 25:1 & 2 (1991): 41–57; Geggus, "Haitian Voodoo in the Eighteenth Century: Language, Culture, and Resistance," *Jahrbuch für Geschichte von Staat, Wirtschaft, und Gesellschaft Latein Amerikas* 28 (1991): 21–51. See also Véronique Campion-Vincent, "L'image du Dahomey dans la presse française (1890–1895): les sacrifices humaines," *Cahiers d'études africaines* 7:25 (1967): 27–58; Robin Law, "Human Sacrifice in Pre-Colonial West Africa," *African Affairs* 84:334 (1985): 53–87; Law, "On the African Background to the Slave Insurrection in Saint-Domingue (Haiti) in 1791: The Bois Caiman Ceremony and the Dahomian 'Blood Pact,'" paper presented at the Harriet Tubman Seminar, Department of History, York University, York, Ontario, 8 November 1999.

24. Cited by R. A. Kea, "Firearms and Warfare on the Gold and the Slave Coasts from the Sixteenth to the Nineteenth Centuries," *Journal of African History* 12:2 (1971): 187.

25. Robert Smith, "The Canoe in West African History," *Journal of African History* 11:4 (1970): 526.

26. M. L. E. Moreau de Saint Méry, *Description topographique, physique, civile, politique, et historique de la partie française de l'île de Saint-Dominque* (Philadelphia: Chez l'Auteur, 1797), 1:25; Hall, *Social Control in Slave Plantation Societies*, 20–21; Gomez, *Exchanging Our Country Marks*, 116–20, 127–28; Gomez, "African Slavery and Identity in the Americas," *Radical History Review* 75 (1999): 111–20.

27. Douglas B. Chambers, "My Own Nation: Igbo Exiles in the Diaspora," *Slavery & Abolition* 18:1 (1997): 75.

28. Daniel A. Offiong, "The Status of Slaves in Igbo and Ibibio in Nigeria," *Phylon* 46:1 (1985): 49.

29. J. C. Cotton, "The People of Old Calabar," *Journal of the Royal African Society* 4:15 (1905): 303.

30. Ibid., 306. See also A. J. H. Latham, "Witchcraft Accusations and Economic Tensions in Pre-Colonial Old Calabar," *Journal of African History* 13:2 (1972): 249–60.

31. Gomez, *Exchanging Our Country Marks*, 127.

32. On the Abakuás, see Roberto Nodal, Rolando A. Alum, and Rafael Núñez, *Linguistic Folklore in the Latin Caribbean: A Selected Glossary of the Abakuá Language in Cuba* (Milwaukee: University of Wisconsin–Milwaukee, 1977); Raphael A. Núñez-Cedeño, Roberto Nodal, and

Rolando A. Alúm, *The Afro-Hispanic Abakuá: A Study of Linguistic Pidginization* (Milwaukee: University of Wisconsin–Milwaukee, 1982); Tato Quiñones, *Ecorie Abakuá: Cuatro ensayos sobre los ñáñigos cubanos* (Havana: Ediciones Unión, 1994); Bárbara Balbuena Gutiérrez, *El íreme Abakuá* (Havana: Pueblo y Educación, 1996).

33. For a firsthand account of the problems of the British with the Ashantee kingdom, see the following editions of *The Times*: 10 July 1816; 8 November 1821; 13 October 1823; and 14 July 1824.

34. See, for example, Gaspar, *Bondmen and Rebels*; Mavis C. Campbell, *The Maroons of Jamaica, 1655–1796: A History of Resistance, Collaboration, and Betrayal* (Granby, Mass.: Bergin & Garvey, 1988); Michael Mullin, *Africa in America: Slave Acculturation in the American South and the British Caribbean, 1736–1831* (Urbana: University of Illinois Press, 1992).

35. Cited in Kea, "Firearms and Warfare," 187.

36. Ibid., 207.

37. Smith, "The Canoe," 517.

38. Gomez, *Exchanging Our Country Marks*, 109.

39. Ibid., 107–10.

40. Sam K. Akesson, "The Akan Concept of the Soul," *African Affairs* 64:257 (1965): 280.

41. Ibid.

42. Beier, "Spirit Children among the Yoruba," 329.

43. Akesson, "The Akan Concept of the Soul," 284.

44. Cited in Gomez, *Exchanging Our Country Marks*, 64.

45. Ibid.

46. Walter Rodney, *A History of the Upper Guinea Coast* (Oxford: Clarendon Press, 1970), 95–113.

47. Willem Bosman, *A New and Accurate Description of the Coast of Guinea* (London: Frank Cass, 1705), 186.

48. See A. P. Kup, *A History of Sierra Leone* (Cambridge: Cambridge University Press, 1962), 171 (quote); Smith, "The Canoe," 525.

49. James L. A. Webb Jr., "The Horse and the Slave Trade between the Western Sahara and Senegambia," *Journal of African History* 34:2 (1993): 229.

50. Thornton, *Africa and the Africans*, 88.

51. See G. Aguirre Beltrán, "The Rivers of Guinea," *Journal of Negro History* 31:3 (1946): 120 (quote); Alessandra Basso Ortiz, "La rumba, ¿Género de origen Gangá?" *La Jiribilla* 65 (2002), footnote 18. Online at: www.lajiribilla.co.cu/2002/n65_agosto/1562_65.html.

52. See Alessandra Basso Ortiz, "Los gangá longobá: el nacimiento de los dioses," *Boletín Antropológico* 2:52 (2001): 196.

53. On the Vai, see Adam Jones, "Who Were the Vai?" *Journal of African History* 22:2 (1981): 159–78. J. M. Harris has written that in the Sierra Leone area, at least in the second part of the nineteenth century, clothes were "made principally in Kissy." Harris, "Some Remarks on the Origin, Manners, Customs, and Superstitions of the Callinas People of Sierra Leone," *Journal of the Anthropological Society of London* 4 (1866): lxxxii.

54. See Aguirre Beltrán, "The Rivers of Guinea," 300.

55. Ibid., 305.

56. Ibid., 298–305.

57. The first comment was offered by the Reverend Abbiel Abbot, who visited Cuba in 1828, while the second opinion belongs to Dolores María de Ximeno. See Basso Ortiz, "La rumba," footnotes 13–15. Online at: www.lajiribilla.co.cu/2002/n65_agosto/1562_65.html.

58. Gangá slaves and their descendants have been frequently linked to the birth of the musical rhythm known as rumba and to the practice of a martial art known as mani. See Fernando Ortiz, Los instrumentos de la música afrocubana (Havana: Cárdenas y Cía, 1954), IV:367; Ortiz, Los bailes y el teatro de los negros en Cuba (Havana: Letras Cubanas, 1981), 427; Basso Ortiz, "La rumba." On Cuban mani, see also Matthias Röhrig Assunção, Capoeira: The History of an Afro-Brazilian Martial Art (London: Routledge, 2005), 62–63.

59. H. C. Creswick, "Life amongst the Veys," Transactions of the Ethnological Society of London 6 (1868): 355.

60. Harris, "Some Remarks," lxxxiii; Creswick, "Life amongst the Veys," 358.

61. Creswick, "Life amongst the Veys," 355, 358.

62. G. F. Scott Elliot, "Some Notes on Native West African Customs," Journal of the Anthropological Institute of Great Britain and Ireland 23 (1894): 82.

63. Creswick, "Life amongst the Veys," 358.

64. Ibid.

65. According to Gwendolyn Midlo Hall, "There were conflicting usages of the term 'Congo' (Kongo) and 'Angola' as coastal terms for Atlantic slave trade voyages as well as for individuals recorded in documents in the Americas." Hall, Slavery and African Ethnicities, 65.

66. On Congo and Angola, see Susan Herlin Broadhead, "Beyond Decline: The Kingdom of the Kongo in the Eighteenth and Nineteenth Centuries," International Journal of African Historical Studies 12:4 (1979): 615–50; John K. Thornton, The Kingdom of Kongo: Civil War and Transition (Madison: University of Wisconsin Press, 1983); Anne Hilton, The Kingdom of Kongo (Oxford: Clarendon Press, 1985). See also James H. Sweet, Recreating Africa: Culture, Kinship, and Religion in the African-Portuguese World, 1441–1770 (Chapel Hill: University of North Carolina Press, 2003).

67. See, for example, Fredrika Bremer, The Homes of the New World: Impressions of America (New York: Harper and Brothers, 1853), letter 35, 1:382–84; Guérin Montilus, "Guinea versus Congo Lands: Aspects of the Collective Memory in Haiti," in Global Dimensions of the African Diaspora, ed. Joseph E. Harris (Washington, D.C.: Howard University Press, 1993), 159–65; Law, "On the African Background," footnote 46; Michael A. Gomez, "African Slavery and Identity in the Americas," Radical History Review 75 (1999): 117–18.

68. Thornton, "African Dimensions of the Stono Rebellion"; Thornton, "'I Am the Subject of the King of Kongo': African Political Ideology and the Haitian Revolution," Journal of World History 4:2 (1993): 181–214; Edison Carneiro, O quilombo dos Palmares, 1630–1695 (São Paulo: Brasiliense, 1947); Décio Freitas, Palmares, a guerra dos escravos (Rio de Janeiro: Graal, 1982); Raymond Kent, "Palmares, An African State in Brazil," Journal of African History 6:2 (1965): 161–75; Mary Karasch, Slave Life in Rio de Janeiro, 1808–1850 (Princeton: Princeton University Press, 1987); Linda M. Heywood, "The Angola-Afro-Brazilian Cultural Connection," Slavery

& Abolition 20:1 (1999): 9–23. See also a recent Brazilian contribution on this issue, João José Reis and Flavio dos Santos Gomes, eds., *Liberdade por um fio: História dos quilombos no Brasil* (São Paulo: Companhia das Letras, 1996).

69. John K. Thornton, "The Art of War in Angola," *Comparative Studies in Society and History* 30:2 (1988): 362. See also Hilton, *The Kingdom of Kongo*, 50–68.

70. Thornton, "The Art of War in Angola," 363.

71. Gomez, *Exchanging Our Country Marks*, 143.

72. After 1800, at least fourteen thousand slaves entered Cuba from Mozambique. See David Eltis, Stephen D. Behrendt, David Richardson, and Herbert Klein, eds., *The Transatlantic Slave Trade: A Database on CD-ROM* (Cambridge: Cambridge University Press, 1999).

73. Hall, *Slavery and African Ethnicities*, 62–64.

74. Cited in Natalia Bolívar, "Los Changaní de Guanabacoa: Reglas de Palo," *Catauro* 2:3 (2001): 211–19.

75. Ibid.

76. Paul Lovejoy, for instance, has suggested that African warfare was the result of political fragmentation. Lovejoy, *Transformations in Slavery*, 66–87.

CHAPTER 2

1. Deposition of Captain Sixto Morejón y Rojas. Criminal proceedings against the rebels of the cafetal Empresa, San Luis de la Ceiba. July 1840. ANC: ME. 595/Ap.

2. Joaquín de Urrutia to the captain-general. San Antonio, 2 July 1840. ANC: GSC. 939/33131. Urrutia accounted for the death of two whites (the owner and the mayoral), one slave killed in the ensuing fight, and seven others who hanged themselves. Nineteen slaves were arrested, two of whom were wounded, and between twenty-five and thirty of whom were still at large.

3. Deposition of Fernando Mandinga. San Luis de la Ceiba. July 1840. ANC: ME. 595/Ap.

4. Genovese, *From Rebellion to Revolution*, 3–4, 35–36, 85–90, 117–25.

5. Captain-General Jerónimo Valdés to the Minister of Ultramar. Havana, 3 September 1843. ANC: AP. 41/60. Some of the most important and biggest slave revolts ever to occur in Cuba took place during Valdés's command. Among them was the terrifying revolt of March 1843 on the plantations around Cárdenas.

6. Although frequently studied together, conspiracies and revolts are far from being the same. Conspiracies are concerted movements that may or may not lead to revolts. A conspiracy involves the agreement of a certain number of individuals who work together to overthrow or change some aspects of the political system under which they live. Because recorded conspiracies often were uncovered before they had the chance to become insurrections, it is very difficult to assess their extent, objectives, or members. Historians usually are obliged to rely upon isolated pieces of evidence and to guess what was happening among the plotters. Revolts, in contrast, are actions that actually happened. Revolts may be either the result of a conspiracy or spontaneous actions or reactions to some specific form of oppression. A revolt is a recorded event about which there are no doubts. A revolt usually leaves behind property damage, deaths, and a long-lasting memory of fear, as happened in the case of Saint Domingue.

Thus when dealing with historical documents relating to revolts, historians have at least a starting point—that the event actually took place. See: Pearson, "Trials and Errors: Denmark Vesey and His Historians"; Thomas J. Davis, "Conspiracy and Credibility: Look Who's Talking about What—Law Talk and Loose Talk"; Robert L. Paquette, "Jacobins of the Lowcountry: The Vesey Plot on Trial," all in William and Mary Quarterly, 3d. ser., 59:1 (2002): 137–42, 167–74, and 185–92. See also Robert L. Paquette and Douglas R. Egerton, "On Facts and Fables: New Light on the Denmark Vesey Affair," South Carolina Historical Magazine 105:1 (2004): 8–48.

7. Paquette, Sugar Is Made with Blood; Rodolfo Sarracino, Inglaterra: sus dos caras en la lucha cubana por la abolición (Havana: Letras Cubanas, 1989).

8. Paquette, Sugar Is Made with Blood, 4.

9. See Manuel Sanguily, José de la Luz y Caballero (estudio crítico) (1890; reprint, Havana: Consejo Nacional de Cultura, 1962); Sanguily, "Un improvisador cubano (el poeta Plácido y el juicio de Menéndez Pelayo)," Hojas Literarias 3 (1894): 93–120; Sanguily, "Otra vez Plácido y Menéndez Pelayo," Hojas Literarias 3 (1894): 227–54; Sanguily, "Una opinión en contra de Plácido (notas críticas)," Hojas Literarias 4 (1894): 425–35; José de Jesús Márquez, "Plácido y los conspiradores de 1844," Revista Cubana 20 (1894): 35–51; Vidal Morales y Morales, Iniciadores y primeros mártires de la revolución cubana (Havana: Consejo Nacional de Cultura, 1963), 147–77.

10. An excellent approach to these discussions can be found in Paquette, Sugar Is Made with Blood, 3–28.

11. Miguel Flores operated both in Havana and Matanzas under the alias of Juan Rodríguez. About his connections with British officials in Cuba, see ibid., 252–56, 264.

12. Deposition of Antonio Lucumí. Gibacoa, May 1844. AOHCH: FG. 117/1.

13. Deposition of Basilio Pérez. Cárdenas, April 1844. ANC: CM. 36/1.

14. Plácido's two most famous poems are "Plegaria a Dios," written just before his execution by a firing squad on the morning of 28 June 1844, and "A mi madre." The debate about La Escalera is far from finished. There are several recent works on Plácido and his role in La Escalera. See Enildo A. García, Cuba: Plácido, poeta mulato de la emancipación, 1809–1844 (New York: Senda Nueva Ediciones, 1986); Walterio Carbonell, "Plácido, ¿Conspirador?" Revolución y Cultura 2 (1987): 57–58; Sarracino, Inglaterra: sus dos caras; Manuel Barcia, "Plácido: Entre la realidad y el mito," Bohemia 91:14 (1999): 64–65.

15. Childs, The 1812 Aponte Rebellion, 120–22.

16. ANC: AP. 12/17. I would like to thank Matt Childs for giving me the opportunity to read such an interesting document, found by him and archivist Jorge Macle after months of searching in the Archivo Nacional de Cuba. A significant part of the explanations given by Aponte about his book of drawings appears in Franco, La conspiración de Aponte, 66–72, 74–97. Franco's work was the most comprehensive analysis of the conspiracy until Childs's book on the subject. See also José de Jesús Márquez, "Conspiración de Aponte," Revista Cubana 19 (1894): 441–54; Angel Augier, "José Antonio Aponte y la conspiración de 1812," Bohemia 54:15 (1962): 48–49, 64; Alain Yacou, "La conspiración de Aponte," Historia y Sociedad 1 (1988): 39–58. Another recent work on Aponte's role in the conspiracy is Stephan Palmié, Wizards and Scientists: Explorations in Afro-Cuban Modernity and Tradition (Durham, N.C.: Duke University Press, 2002), 79–158.

17. Childs, *The 1812 Aponte Rebellion*, 163–68. Curiously, thirty-six years later, during the interrogations that followed the discovery of La Escalera, a Creole slave from the sugar mill La Emilia in Matanzas declared that he had heard some time before that "the negroes from Guarico had won the war, and that consequently Jean François had come to Havana from Saint Domingue." Deposition of Juan Criollo. April 1844. ANC: CM. 36/1.

18. Childs, *The 1812 Aponte Rebellion*, 121–23.

19. This conspiracy has not been seriously studied. The best approaches to it can be found in Gloria García, "A propósito de La Escalera: El esclavo como sujeto politico," *Boletín del Archivo Nacional de Cuba* 12 (2000): 1–13; Ada Ferrer, "Noticias de Haití en Cuba," *Revista de Indias* 63:229 (2003): 675–94.

20. Deposition of Estanislao, slave native to Guarico (French Saint-Domingue). Güines, May 1806. ANC: AP. 9/27.

21. The term contramayoral—Spanish for slave driver—was widely used in colonial Cuba to designate the immediate subaltern of the mayoral or mayoral. Most of the time the job was performed either by free blacks and mulattos or by slaves. See Robert L. Paquette, "The Drivers Shall Lead Them: Image and Reality in Slave Resistance," in *Slavery, Secession, and Southern History*, ed. Robert L. Paquette and Louis Ferleger (Charlottesville: University Press of Virginia, 2000), 31–58; Manuel Barcia, "Un aspecto de las relaciones de dominación en la plantación esclavista cubana: Los contramayorales esclavos," *Gabinete de Arqueología: Boletín* 1:2 (2001): 88–93.

22. Deposition of Francisco Joseph Fuertes. Güines, May 1806. ANC: AP. 9/27.

23. Geggus, "Slave Resistance in the Spanish Caribbean," 133; Alain Yacou, "La présence française dans la partie occidentale de l'île de Cuba au lendemain de la Révolution de Saint-Domingue," *Revue Française d'Histoire d'Outre-Mer* 84 (1987): 149–88.

24. Intervention of Luis de las Casas in the Junta of the Real Consulado de Agricultura y Comercio held on 18 November 1795. ANC: RCJF. Book 161. See also Childs, *The 1812 Aponte Rebellion*, 38–39.

25. Geggus, "Slave Resistance," 133–34.

26. Ibid., 138.

27. ANC: AP. 7/30. The plot was uncovered in April, when the slave Guillermo from the ingenio Buenavista told his mayoral about the conspiracy. In October 1798 another outbreak took place on the Calvo estate near Güines. Geggus, "Slave Resistance," 151.

28. See García, *La esclavitud*, 112–28.

29. Deposition of José Gangá. Ingenio Unión, Guamutas, April 1844. ANC: CM. 36/1. Jacobo was allowed access to his master's food because he was the cook for the sugar mill. See also the testimonies offered by José Lucumí, José María Congo, and the free black Pancho Peraza, all included in the same file.

30. Depositions of Lino, Canuto Criollo, and Camilo Macuá. Guamutas, April 1844. ANC: CM. 36/1.

31. Criminal proceedings against Pedro José Mandinga for killing the manager Don José Cartas. Jibacoa, January 1824. ANC: ME. 813/F.

32. Criminal proceedings against Mariano Gangá and Severino Lucumí for killing their owner, Don José Delgado. Cimarrones, November 1832. ANC: ME. 481/C.

33. Deposition of Pedro Gangá. Ingenio Unión, Guamutas, April 1844. ANC: CM. 36/1.

34. Criminal proceedings against Luis Lucumí and Esteban Mina for battering their mayoral, Juan Camejo. Pinar del Río. September 1826. AHPPR: IJC. 34/73.

35. Proceedings against Gervasio Carabalí for criminal behavior. Bainoa. April–June 1822. ANC: ME. 679/F.

36. Deposition of Rafael Gangá. Catalina de Güines, September 1827. ANC: ME. 1069/B.

37. On warfare and firearms in West and West-Central Africa, see Thornton, "The Art of War in Angola"; Thornton, Africa and the Africans; Thornton, Warfare in Atlantic Africa, 1500–1800 (London: Routledge, 1999); Webb, "The Horse and Slave Trade"; Law, "A West African Cavalry State"; Kea, "Firearms and Warfare on the Gold and Slave Coasts," 185–213.

38. Geggus, "Slave Resistance," 138.

39. I refer here to the slave uprisings that occurred on the cafetal La Esperanza in 1817 and in the sugar district of El Mariel on 3 May 1822. Diario del Gobierno de la Habana 68 (9 March 1817): 1–3; ANC: AP. 20/17.

40. On this uprising, see Manuel Barcia, La sublevacion de esclavos de 1825 en Guamacaro (M.A. thesis, University of Havana, 2000).

41. Reglamento de policía rural de la jurisdicción del Gobierno de Matanzas (1825). ANC: GSC. 1469/57999.

42. Francisco Dionisio Vives, Circular á toda la Isla y Capitanías de esta jurisdicción (Havana, 13 July 1830). AHPM: GP. 7/4.

43. Bozen Mandinga Moro is probably the only certain reference to a Muslim slave in Cuba. Since the early fifteenth century, the term "Moro" was broadly used by Spanish-speaking peoples in reference to believers in Islam. Bozen's ethnic and geographic origin also are also suggestive of Muslim beliefs, since Islam was well-known among Mandingas at the time. Bozen was thirty-five years old in 1830. The region of Guamacaro was also known as Limonar. ANC: ME. 713/B.

44. Wenceslao de Villaurrutia to Jerónimo Valdés. Havana, 25 March 1842. ANC: GSC. 940/33158.

45. See the confession of Pablo Gangá in 1844. ANC: CM. 43/1.

46. Count of Fernandina, Antonio Bermúdez de Castro, and Francisco Fesser to Colonel Rafael O'Farrill y Herrera. Havana, 11 January 1827. ANC: RCJF. 150/7436.

47. Lombillo's concerns were transmitted by Colonel Miranda y Madariaga to the captain-general. Colonel Joaquín Miranda y Madariaga to Captain-General Francisco Dionisio Vives. Cafetal Reunión, 8 January 1827. ANC: RCJF. 150/7436.

48. Proceedings against the insubordinate slaves from the cafetal El Carmen. Güira de Melena, October–November 1827. ANC: ME. 223/F.

49. Deposition of Simón Mina. November 1827. ANC: ME. 223/F.

50. Criminal proceedings against seventeen slaves from the sugar mill Purísima Concepción. Sibanacán, Güira de Melena, September 1832. ANC: ME. 570/S.

51. Cholera morbus had a reputation of being a disease of blacks, since apparently Africans and African descendants were far more susceptible to contracting and dying from it. See Kenneth Kiple and Virginia Himmelsteib King, Another Dimension to the African Diaspora: Diet, Disease, and Racism (Cambridge: Cambridge University Press, 1981), 147–57; Kenneth

Kiple, *The Caribbean Slave: A Biological History* (Cambridge: Cambridge University Press, 1984), 135, 146–48; Kiple, "Cholera and Race in the Caribbean," *Journal of Latin American Studies* 17:1 (1985): 157–77; Adrián López Denis, *El cólera en la Habana: La epidemia de 1833* (M.A. thesis, University of Havana, 2000); López Denis, "Higiene pública contra higiene privada: cólera, limpieza y poder en La Habana colonial," *Estudios Interdisciplinarios de América Latina y el Caribe* (EIAL) 14:1 (2003): 11–34.

52. This revolt is one of the few that has been subject of at least one article. See Juan Iduarte, "Noticias sobre sublevaciones y conspiraciones de esclavos. Cafetal Salvador, 1833," *Revista de la Biblioteca Nacional José Martí* 73:24 (1982): 117–52.

53. Deposition of Diego Barreiro. Cafetal Salvador, August 1833. Barreiro, who was badly injured in one of his arms, declared that one of the old slaves, Alejo Lucumí, was able to tell him in Spanish that the slaves just wanted their freedom. He said, "Carajo, queremos libertad." ANC: ME. 540/B.

54. Deposition of Matías Eguyoví Lucumí. Cafetal Salvador, August 1833. Eguyoví later declared that the slaves were playing the same drums in the revolt that they used to play at their parties. ANC: ME. 540/B.

55. All quotations are from ANC: ME. 540/B.

56. Depositions of Matías Eguyoví Lucumí and Pacual Ayaí Lucumí. Cafetal Salvador, August 1833. ANC: ME. 540/B.

57. Deposition of the mayoral, Don Manuel Díaz. Bauta, 3 August 1834. ANC: ME. 451/F.

58. Deposition of the contramayoral, Tomás Lucumí. Bauta, 4 August 1834. ANC: ME. 451/F.

59. Depositions of Antonino Lucumí and contramayoral Tomás Lucumí. Bauta, 4 August 1834. ANC: ME. 451/F.

60. Proceedings against the rebel Lucumí slaves of the ingenio La Magdalena. Jurisdiction of Santa Ana, Matanzas. July 1835. ANC: ME. 232/Z. The events of La Magdalena are extensively examined in the Chapter 4 of this study.

61. Deposition of Valentín Lucumí. Ingenio La Sonora. June, 1837. ANC: ME. 1178/B.

62. Deposition of Fermín Lucumí. Ingenio La Sonora. June 1837. ANC: ME. 1178/B.

63. Sentence dictated by the Military Commission against the rebel slaves from the ingenio La Sonora. Havana, October 1837. Some of the slaves of La Sonora were taken to witness the execution of their comrades in Havana.

64. Statement of Lieutenant José Vicente Torro, defense attorney of Esteban Lucumí. Havana, 14 October 1837. ANC: ME. 1178/B.

65. Proceedings against the slaves of the sugar mill San Pablo, owned by Don Julián Zaldivar. Catalina de Güines, September 1837. ANC: ME. 1193/H.

66. Ibid.

67. Inocencio López Gavilán to the captain-general. San Nicolás, 15 June 1840. ANC: GSC. 939/33130.

68. In 1836, Lucumí slaves had revolted in Havana in a sawmill located in the suburb of El Cerro, less than forty minutes by foot from the center of the capital. Proceedings against the Lucumí rebels from the sawmill located in Cruz del Padre street, owned by Don Juan Echarte, 1836. ANC: ME. 731/A.

69. Depositions of Valentín Toledo and Domingo Aldama. Havana, October 1841. ANC: GSC. 940/33154.

70. Proceedings against 26 Lucumí, 11 Arará, 4 Mina and 1 Congo slaves from the sugar mill La Arratia. July 1842. ANC: CM. 28/1.

71. Joseph T. Crawford to the earl of Aberdeen. No. 15, Havana, 18 April 1843. TNA: FO. 72/634.

72. Ibid.

73. British and Foreign Anti-Slavery Reporter, 31 May 1843, 82. See also John G. Wurdermann, Notas sobre Cuba (Havana: Letras Cubanas, 1989): 272–73; Paquette, Sugar Is Made with Blood, 177–78. The most recent study on this movement is Daniel Martínez Garcia, "La sublevación de la Alcancía: su rehabilitación histórica en el proceso conspirativo que concluye en La Escalera (1844)," Rábida 19 (2000): 41–48.

74. Wurdermann, Notes, 271–72; Paquette, Sugar Is Made with Blood, 178.

75. Diario de la Habana (8 November 1843), 1–2; Paquette, Sugar Is Made with Blood, 209–15. Polonia Gangá, a slave from the plantation Santísima Trinidad de Oviedo, denounced the conspiracy. See María del Carmen Barcia and Manuel Barcia, "La conspiración de La Escalera: el precio de una traición," Catauro 2:3 (2001): 199–204.

76. See K. G. Davies, The Royal African Company (London: Longmans, Greens, 1957); Jean Meyer, L'Armement nantais dans la deuxième moitié du XVIIIe siècle (Paris: SEVPEN, 1969); Kwame Yeboah Daaku, Trade and Politics on the Gold Coast, 1600–1720 (Oxford: Clarendon Press, 1970); Gavin White, "Firearms in Africa," Journal of African History 12:2 (1971): 173–84; Philip D. Curtin, Economic Change in Precolonial Africa (Madison: University of Wisconsin Press, 1975); Joseph. E. Inikori, "The Import of Firearms into West Africa, 1750–1807," Journal of African History 18:3 (1977): 339–68; David Richardson, "West African Consumption Patterns and Their Influence on the Eighteenth-Century English Slave Trade," in The Uncommon Market: Essays in the Economic History of the Atlantic Slave Trade, ed. Henry. A. Gemery and Jan. S. Hogendorn (New York: Academic Press, 1979); W. A. Richards, "The Import of Firearms into West Africa in the Eighteenth Century," Journal of African History 21:1 (1980): 43–59.

77. Deposition of Martín Gangá. Ingenio Amunátegui, July 1844. ANC: CM. 36/1.

78. Deposition of Perico Criollo. Matanzas, January 1844. ANC: CM. 37/1.

79. Deposition of Alejandro Gangá. Ingenio Coto, Matanzas. January 1844. ANC: CM. 37/1.

80. Deposition of Tomás Criollo. Ingenio Coto, Matanzas. January 1844. ANC: CM. 37/1.

81. There are several published works about "African" oaths in the New World. See, for example, Barbara K. Kopytoff, "Colonial Treaty as Sacred Charter of the Jamaican Maroons," Ethnohistory 26 (1979): 45–64; Kenneth Bilby, "Swearing by the Past, Swearing to the Future: Sacred Oaths, Alliances, and Treaties among the Guianese and Jamaican Maroons," Ethnohistory 44:4 (1997): 655–89.

82. Deposition of José Gangá. Gibacoa, May 1844. AOHCH: FG. 117/1.

83. Depositions of Juliano Gangá, Santiago Congo, and Higinio Mozambique. Gibacoa, May 1844. AOHCH: FG. 117/1.

84. Deposition of Domingo Lucumí. Matanzas, March 1844. ANC: CM. 35/4.

85. For the revolt of Guaycanamar, see José María André to the captain-general, Prince of Anglona. Jaruco, 16 September 1840. ANC: AP. 136/8.

86. Deposition of Rafael Gangá. Catalina de Güines, September 1827. ANC: ME. 1069/B.

87. For the Carabalís of La Hermita, see ANC: CM. 1/3. For the Congos of Cayajabos, see Deposition of José Angel Congo. Cayajabos, November 1812. ANC: ME. 743/Ll.

88. Deposition of Francisco Congo. Gibacoa, May 1844. AOHCH: FG. 117/1.

89. José Martínez to the governor of Matanzas. Limonar, 15 June 1825. ANC: CM. 1/3.

90. Depositions of Diego Barreiro and Pascual Ayaí Lucumí. Cafetal Salvador. August 1833. ANC: ME. 540/B.

91. Deposition of Roque Gangá. Ingenio La Mercedita. April 1844. ANC: CM. 36/1.

92. Deposition of Gertrudis Carabalí. Ingenio La Mercedita. April 1844. ANC: CM. 36/1.

93. Thornton, *Africa and the Africans*, 184–92. According to Clarke, in 1875, "Hausa, Fulah and Mandingo" all served as lingua franca among the Ashantees. Hyde Clarke, "On the Relations of Culture of the Ashantees," *Journal of the Anthropological Institute of Great Britain and Ireland* 4 (1875): 122.

94. Deposition of Tomás Criollo. Ingenio Coto. January 1844. ANC: CM. 37/1.

95. The rebellions of 1825 and 1843 are examples of this pattern.

96. Deposition of José Lucumí. Ingenio Toro. February 1844. ANC: ME. 37/1.

97. Deposition of Antonio Criollo. Ingenio Saratoga. January 1844. ANC: ME. 37/1.

98. Deposition of Tomás Criollo. Ingenio Adan. January 1844. ANC: ME. 37/1.

99. For this debate, see Chapter 4.

100. Gala was referring to Africa. Final conclusions given by prosecutor Apolinar de la Gala. October 1844. ANC: CM. 38/1.

101. Ibid.

102. See, for example, Robin Law, "Human Sacrifice in Pre-Colonial West Africa," *African Affairs* 84:334 (1985): 53–87.

103. Barcia and Barcia, "La conspiración de la Escalera," 200–01.

104. Ibid.

105. Usually this process of renaming—in which the last recognizable African feature was erased—took time and changed step by step. Pablo Gangá, a leader in the 1825 revolt of Guamacaro, was known in 1844 as Pablo Linares Gangá. The Spanish surname "Linares" was taken from his owner, Don Juan Linares. In 1825, approximately four years after his arrival in Cuba, Pablo was referred to as Pablo Gangá de Tosca. Here, the name of his master was also attached to his own name, but the master's name appeared as a reference and not as part of his surname. Confession of Pablo Gangá. Matanzas, 1844. ANC: CM. 43/1.

106. ANC: CM. 70/1 and 2. See also Paquette, *Sugar Is Made with Blood*, 67–68, 264–65.

107. Childs, *The 1812 Aponte Rebellion*, 123–28.

CHAPTER 3

1. José del Mazo to Antonio García Oña. Ingenio La Esperanza, 21 April 1844. AHPM: GP. 8/115.

2. See Reis and Gomes, eds., *Liberdade por um fio*.

3. See Miguel Barnet, ed., *The Autobiography of a Runaway Slave: Esteban Montejo*, trans. Jocasta Innes (New York: Pantheon, 1968); Alex Haley, *Roots* (London: Hutchinson, 1977).

4. See, for example, L. A. Eyre, "The Maroon Wars in Jamaica: A Geographical Appraisal," *Jamaican Historical Review* 12 (1980): 80–102; David Barker and Balfour Spence, "Afro-Caribbean Agriculture: A Jamaican Maroon Community in Transition," *Geographical Journal* 154:2 (1988): 198–208. For agricultural technologies in West Africa, see: M. J. Mortimore, "Population Densities and Systems of Agricultural Land Use in Northern Nigeria," *Nigerian Geographical Journal* 14 (1971): 3–15; G. Benneh, "Systems of Agriculture in Tropical Africa," *Economical Geography* 48 (1972): 244–57; P. Richards, *Indigenous Agricultural Revolution: Ecology and Food Production in West Africa* (London: Hutchinson, 1985).

5. Gabino La Rosa Corzo, *Los cimarrones de Cuba* (Havana: Ciencias Sociales, 1988); La Rosa Corzo, *Runaway Slave Settlements in Cuba.*

6. Luisa R. R. Volpato, "Quilombos in Matto Grosso: Resistência negra em área de fronteira," in *Liberdade por um fio*, ed. Reis and Gomes, 213–39; Mary Karasch, "Os quilombos do ouro na capitania de Goiás," ibid., 240–62; Flavio dos Santos Gomes, "Quilombos do Rio de Janeiro no século XIX," ibid., 263–90.

7. Matthias Röhrig Assunção, "Quilombos maranhaenses," in *Liberdade por um fio*, ed. Reis and Gomes, 433–66; Wim Hoogbergen, *The Boni Maroon Wars in Suriname* (Leiden: E. J. Brill, 1990).

8. João José Reis, "Escravos e coiteiros no quilombo do Oitizeiro," in *Liberdade por um fio*, ed. Reis and Gomes, 332–72.

9. Stuart B. Schwartz, "Cantos e quilombos numa conspiração de escravos haussás," in *Liberdade por um fio*, ed. Reis and Gomes, 373–406; Reis, *Slave Rebellion in Brazil*, 49, 55–57; Manuel Barcia, *Slave Rebellions*, 60–3; Pedro Deschamps Chapeaux, "Etnias africanas en las sublevaciones de los esclavos en Cuba," *Revista Cubana de Ciencias Sociales* 10:4 (Jan.–Apr. 1986): 14–17.

10. Richard Price, *Maroon Societies: Rebel Slave Communities in the America.* (Garden City, N.Y.: Anchor Press, 1973); Price, *The Guiana Maroons: A Historical and Bibliographical Introduction* (Baltimore: Johns Hopkins University Press, 1976).

11. See: Price, *Maroon Societies*; Sidney W. Mintz and Richard Price, *An Anthropological Approach to the Afro-American Past: A Caribbean Perspective* (Philadelphia: ISHI, 1976); Campbell, *The Maroons of Jamaica, 1655–1796*; Hoogbergen, *The Boni Maroon Wars.*

12. Especially noteworthy, and full of inaccuracies, is Miguel Barnet, ed., *The Autobiography.* See also the critiques of Michael Zeuske, "The Cimarrón in the Archives: A Re-Reading of Miguel Barnet's Biography of Esteban Montejo," *New West Indian Guide/Nieuwe West-Indische Gids* 71: 3–4 (1997): 265–79; Zeuske, "Novedades de Esteban Montejo," *Revista de Indias* 59: 216 (1999): 521–25; Zeuske, "Más novedades de Esteban Montejo," *Del Caribe* 38 (2002): 95–101.

13. Ramiro Guerra, *Manual de historia de Cuba* (Havana: Ciencias Sociales, 1971), 483.

14. Deschamps Chapeaux, "Etnias africanas," 14–17.

15. La Rosa Corzo, *Los cimarrones* and *Runaway Slave Settlements.*

16. *Nuevo reglamento y arancel que debe gobernar en la captura de esclavos cimarrones, aprobado por S. M. en Real Orden expedida en San Lorenzo con fecha de veinte de diciembre de 1796* (1796). Like most of the important documents of the time, the code was the work of Francisco de Arango y Parreño, the most influential member of the royal council. I thank Matt Childs for providing me with a copy of this document.

17. Robert Charles Dallas, *The History of the Maroons, from Their Origin to the Establishment of Their Chief Tribe at Sierra Leone: Including the Expedition to Cuba, for the Purpose of Procuring Spanish Chasseurs; and the State of the Island of Jamaica for the Last Ten Years, with a Succinct History of the Island Previous to That Period*. (London: T. N. Longman and O. Rees, 1803), 2:3–109.

18. See the censuses of 1776, 1792, 1817, and 1827. In 1827, the number of slaves and free blacks peaked at 393,436, while whites accounted for 311,051 individuals.

19. *Nuevo reglamento*, 1, 6.

20. Ibid., 9–14.

21. Ibid., 4.

22. Francisco de Arango y Parreño, *De la factoría a la colonia* (Havana: Publicaciones de la Secretaría de Educación La Habana, 1936), 21–113.

23. Records of the ayuntamiento de la Habana corresponding to 18 August 1819. AOHCH: ACT. 1819. Book 96, f. 277vt.

24. Brigadier José Eusebio Escudero, *Reflexiones que sobre los palenques de negros cimarrones respecto a la parte oriental de esta Ysla de Cuba, se elevaron al conocimiento de S.M.* The original document was drawn up in 1816; the copy I cite here was printed in 1819. ANC: RCJF. 141/6935.

25. Captain-General Nicolás Mahy to the captain of the jurisdiction of Cayajabos. Havana, May 1822. ANC: AP. 124/99. There are many other recorded cases of maroon groups assaulting people and stealing goods throughout "civilized areas." See, for instance, the cases of Sabanilla del Encomendador in 1832, AHPM: GP. 7/11; Boca de Jaruco in 1836, ANC: ME. 591/V; and Batabanó in 1842, ANC: GSC. 941/33193.

26. Records of the ayuntamiento de la Havana corresponding to 18 August 1819. AOHCH: ACT. 1819. Book 96, fs. 278–79.

27. Records of the ayuntamiento de la Havana corresponding to 18 August 1819. AOHCH: ACT. 1819. Book 96, f. 278 vt.

28. Gaspar Antonio Rodríguez to Captain-General Nicolás Mahy. 18 June 1822. ANC: RCJF. 150/7416.

29. Journal of José Pérez Sánchez. October 1828. ANC: RCJF. 150/7445.

30. There are also references to a "fresh supply of bloodhounds" shipped to Florida in 1840 to chase either Indians or slaves. *British and Foreign Anti-Slavery Reporter*, 20 May 1840, 111.

31. Dallas, *History of the Maroons*, 2:1–136; Records of the ayuntamiento de la Havana corresponding to 18 August 1819. AHCH: ACT. 1819. Book 96, f. 279.

32. Dallas, *History of the Maroons*, 2:4–5.

33. Ibid.

34. Ibid., 2:58–59.

35. For example, rancheador José Pérez Sánchez used to return on the first day of every month to his house and headquarters in the Loma del Rubí to rest, to supply his men with food for the next month, and to check his orders. On the fifth day, he and his men left for the mountains to begin another twenty-five-day hunting cycle. See the journals of José Pérez Sánchez. ANC: RCJF. 142/6949, 150/7445, 150/7447, 150/7450, 150/7456, 150/7458.

36. Journal of Gaspar Antonio Rodríguez, especially the entries between August and November 1820. ANC: RCJF. 141/6940.

37. Journal of Matías Pérez Sánchez. Entry of 31 March 1815. ANC: RCJF. 141/6925.

38. Journal of Gaspar Antonio Rodríguez. Entry of 31 July 1820. ANC: RCJF. 141/6940.

39. Ibid. Entries between 4 October and 17 November 1820.

40. Journal of Fernando de Osuna. January 1833. ANC: RCJF. 142/6951.

41. Journal of José Pérez Sánchez. El Rubí, 30 September 1828. ANC: RCJF. 142/6949.

42. Ibid. Cayajabos, 23 April 1828. ANC: RCJF. 142/6949.

43. Journal of Gaspar Antonio Rodríguez. July–November 1820. ANC: RCJF. 141/6940.

44. Dallas, History of the Maroons, 2:61.

45. Ibid., 2:64–65.

46. Ibid., 2:56–57.

47. Ibid., 2:63–64.

48. Ibid., 2:62–63.

49. Ibid., 2:57–58.

50. María de las Mercedes Santa Cruz y Montalvo, countess of Merlin, Viaje a La Habana (Madrid: Sociedad Literaria, 1841), 31–33. Dallas also shared the countess of Merlin's opinion, although he did not mention the attack to a maroon's ears. Rather, after affirming that the dogs "will not kill the object they pursue unless resisted," he wrote that "on coming up with a fugitive, they bark at him till he stops, they then couch near him, terrifying him with a ferocious growling of the stirs. In this position they continue barking to give notice to the chasseurs, who come up and secure their prisoner." Dallas, History of the Maroons, 2:56–57.

51. Journal of José Pérez Sánchez. January 1828. ANC: RCJF. 142/6949.

52. Ibid. November 1828.

53. Ibid. January 1831. ANC: RCJF. 150/7458.

54. José Borroto to the governor of Matanzas. Camarioca, 21 April 1844. AHPM: GP. 12/22.

55. Journal of José Pérez Sánchez. April 1828. ANC: RCJF. 142/6949.

56. Joaquín Figueredo, Lieutenant of Canasí, to the governor of Matanzas. Canasí, 2 July 1832. AHPM: GP. 7/7.

57. The impact of runaway slaves in the rural landscape was reflected in many ways, including toponymic changes. On 19 April 1831, José Pérez Sánchez and his men rested for a nightin a place known as "Derrumba Piedra" or "Falling Rocks." He wrote in his journal that the place had received this curious name from the "Negroes." ANC: RCJF. 150/7458.

58. The captain of the jurisdiction of Cimarrones to Captain-General Jerónimo Valdés, Cimarrones, 2 March 1841. ANC: RCJF. 142/7008.

59. Antonio María de la Torre to the prior and síndicos del ayuntamiento de Remedios. Remedios, 4 February 1831. ANC: RCJF. 150/7459.

60. Esteban Montejo is perhaps the most notorious example of a maroon who was said to have lived alone in the woods for years. His real story, however, is likely to have been quite different from the tale he told Miguel Barnet in the early 1960s. See Zeuske, "The Cimarrón," "Novedades," and "Más novedades."

61. In 1816, Lieutenant of Infantry Manuel de Chenard defined palenques in a simple sentence: "The name of palenques has been given to those caves or woods where the slaves find refuge and gather together with the main purpose of getting rid of the works of their

owners." Manuel de Chenard, *Ynstrucción para Govierno del Tente de Ynfanta de la Habana*. Cuba, 1816. AHPSC: GP. 554/1.

62. Journal of Matías Pérez Sánchez. March 1825. ANC: RCJF. 141/6925.

63. Chenard, *Ynstrucción*. AHPSC: GP. 554/1.

64. Vultures were often the best guides to palenques for rancheadores, and there are many references to them in the rancheadores' journals. Fernando de Osuna revealed the extent of rancheadores' trust that vultures would lead them to maroons, writing in late 1833, "We saw some vultures near the cliff, which indicates the place where the black maroons are." Journal of Fernando de Osuna. December 1833. ANC: RCJF. 142/6951.

65. On warfare in Africa, see Thornton, "The Art of War in Angola"; Thornton, *Africa and the Africans*; Thornton, *Warfare in Atlantic Africa*; Webb, "The Horse," 221–46; Law, "A West African Cavalry State"; Kea, "Firearms and Warfare."

66. Francisco Chappottín to Captain-General Francisco Dionisio Vives. Cayajabos, 15 January 1829. ANC: RCJF. 150/7447.

67. Journal of Fernando de Osuna. January 1833. ANC: RCJF. 142/6951.

68. Chenard, *Ynstrucción*.

69. Pancho Mina, a maroon on the run for more than twenty years, declared that in his palenque there "was no captain among them, that everyone governs itself individually." Deposition of Pancho Mina. Cayajabos, 31 August 1835. ANC: GSC. 616/19688.

70. Mandinga was finally captured in April and executed in May 1828. Prosecutor Francisco Seidel wrote about Mandinga's reputation among the slaves of being impossible to capture and recommended cutting off his head and his remaining hand, exhibiting them "in the most public place of the main road until the time will consume them." Recommendation of Prosecutor Francisco Seidel. Havana, May 1828. ANC: C.M. 2/4, fs. 47–57vt.

71. Journal of José Pérez Sánchez. January 1831. ANC: RCJF. 150/7458.

72. Records of the Real Consulado de Agricultura y Comercio de la Habana. Board of 21 January 1829. ANC: RCJF. 150/7447.

73. About this issue in other areas of the Americas, see Michael Craton, "Proto-Peasant Revolts? The Late Slave Rebellions in the British West Indies, 1816–1832," *Past and Present* 85 (1979): 110–11; Barker and Spence, "Afro-Caribbean Agriculture."

74. Deposition of Francisco (Pancho) Mina. Cayajabos, 31 August 1835. ANC: GSC. 616/19688.

75. Deposition of Rita Bibí. Santiago de Cuba, August 1838. ANC: ME. 1107/F.

76. Deposition of Francisco (Pancho) Mina. Cayajabos, 31 August 1835. ANC: GSC. 616/19688.

77. Deposition of Dionisia Calás Gangá. Santiago de Cuba, August 1838. ANC: ME. 1107/F.

78. Deposition of Federico Bibí. Santiago de Cuba, August 1838. ANC: ME. 1107/F.

79. Deposition of Manuela Calás Gangá. Santiago de Cuba, August 1838. ANC: ME. 1107/F.

80. In fact, maroons frequently established their palenques in places where they found wild yams or fruit trees, as in the case of the ranchería founded by Bartolomé Portuondo

Congo and his companions in Demajagual. Deposition of Bartolomé Portuondo Congo. Santiago de Cuba, August 1838. ANC: ME. 1107/F.

81. Eusebio Escudero to the Real Consulado. Santiago de Cuba, 13 February 1819. ANC: RCJF. 141/6935.

82. Deposition of Francisco (Pancho) Mina. Cayajabos, 31 August 1835. ANC: GSC. 616/19688.

83. Journal of Gaspar Antonio Rodríguez. July 1820. ANC: RCJF. 141/6940.

84. Journal of José Pérez Sánchez. October 1828. ANC: RCJF. 150/7445.

85. [Francisco Estévez], Diario del rancheador, ed. Cirilo Villaverde (Havana: Letras Cubanas, 1982), 127.

86. Journal of Gaspar Antonio Rodríguez. August 1820. ANC: RCJF. 141/6940.

87. Gaspar Antonio Rodríguez to Captain-General Mahy. 1 March 1822. ANC: RCJF. 150/7416. Referring to the sugar mill San Roque, owned by Xavier Pedroso, Rodríguez also stated, "The maroons are tolerated on this estate in a way that whoever sees them, thinks that they are not maroons."

88. José Pérez Sánchez, for example, reported killed cattle in the estate El Brujo in September 1828. Journal of José Pérez Sánchez. October 1828. Soon afterwards Pérez Sánchez also reported plunderings in banana fields on the plantations San Luis (October 1828) and Pluma (January 1829). ANC: RCJF. 142/6949, 150/7445.

89. Joaquín Ramos to the governor of Matanzas. Sabanilla del Encomendador, 22 February 1832. AHPM: GP. 7/11.

90. Journal of José Pérez Sánchez. January, 1831. ANC: RCJF. 150/7458.

91. Deposition of Domingo López. Cayajabos, 1836. ANC ME 563/A. This deposition was backed by the testimony of José María Díaz, owner of the cafetal San Julián, located in the same area.

92. Joaquín Figueredo, Lieutenant of Canasí, to the governor of Matanzas. Canasí, 2 July 1832. AHPM: GP. 7/7.

93. José Garcilaso de la Vega to Captain-General José Cienfuegos. Cayajabos, 30 January 1819. ANC: RCJF. 141/6934.

94. Journal of José Pérez Sánchez. August 1830. ANC: RCJF. 150/7456.

95. Gaspar Antonio Rodríguez to Captain-General Mahy. 1 March 1822. ANC: RCJF. 150/7416.

96. Eusebio Escudero to the Real Consulado. Santiago de Cuba, 13 February 1819. ANC: RCJF. 141/6935. This figure should be taken with extreme prudence, since the slave population of the area was not very large at the time. According to the 1817 census, the slave population of the entire jurisdiction of Baracoa was 1,850 individuals. The three hundred maroons said to be living in El Frijol one year earlier would have comprised a staggering 16.21 percent of the slave population of Baracoa. See Cuadro estadístico de la siempre fiel Isla de Cuba, correspondiente al año 1827 (Havana: Imprenta del Gobierno, 1818).

97. Deposition of Mateo Lucumí. Ingenio La Andrea, Sabanilla, January 1844. Mateo recounted how his companions Marcelino and Marcos had instructed him to talk with the maroon gang of José Dolores in order to be able to count on him once the uprising had be-

gan. ANC: CM. 37/1. See also Juan Sánchez, "José Dolores, capitán de cimarrones. Un capítulo
inédito de las rebeldías de esclavos en Matanzas," *Bohemia* 66 (November 1974): 50–53.

98. Although the presence of maroons and maroon communities had some institutional
consequences, they never really challenged the very existence of colonial rule in Cuba. See La
Rosa Corzo, *Los cimarrones.*

CHAPTER 4

1. However, some scholars argue that suicide was an unlikely event among the inhabitants
of the Niger Delta. See Gomez, *Exchanging Our Country Marks,* 117.

2. See Toyin Falola and Michael R. Doortmont, "Iwe Itan Oyo: A Traditional Yoruba History
and Its Author," *Journal of African History* 30:2 (1989): 301–29. Suicide, at least as a death penalty,
was a common practice throughout the history of the Oyo. The Oyo Mesi, a sort of supreme
council of state, had the right to condemn any deficient or inept Oba to commit suicide. See
Michael Crowder, *The Story of Nigeria* (London: Faber and Faber, 1962), particularly the first
chapter. For particular cases of suicide among the Yoruba between 1852 and 1885, see J. D. Y. Peel,
Religious Encounters and the Making of the Yoruba (Bloomington: Indiana University Press, 2000).

3. Toyin Falola, *Yoruba Gurus: Indigenous Production of Knowledge in Africa* (Trenton, N.J.:
Africa World Press, 1999), 289.

4. See João José Reis, *Death Is a Festival: Funeral Rites and Rebellion in Nineteenth-Century
Brazil* (Chapel Hill: University of North Carolina Press, 2003), 145.

5. See Robert L. Stein, *The French Slave Trade in the Eighteenth Century: An Old Regime Business*
(Madison: University of Wisconsin Press, 1979), 73–95; Robin Law, *The Slave Coast of West Africa,
1550–1750: The Impact of the Atlantic Slave Trade on an African Society* (Oxford: Clarendon Press, 1991).

6. Jacques Savary, *Le parfait négociant* (Lyon: Jacques Lyons, 1697), 206.

7. Felipe Nery to Antônio Estes de Acosta. Pernambuco, 6 August 1812. Cited in Schwartz,
Sugar Plantations in the Formation of Brazilian Society, 370.

8. There are no references about suicide practices among African Islamized slaves, we
can safely assume that suicide rates were low. The Malikiti Sharia law, adopted in West Africa
since the twelfth century, forbids suicide. The Koran itself has some dispositions regulating
this behavior. See sura 2, verse 195: "Do not with your own hands cast yourselves into destruction"; and sura 4, verse 29: "Do not kill yourselves. God is merciful to you, but he that does
that through wickedness and injustice shall be burned in fire." I thank Robin Law and John
Thornton for their insightful comments and recommendations regarding this issue.

9. James Pope-Hennessy, *Sins of the Fathers: A Study of the Atlantic Slave Traders, 1441–1807*
(London: Weidenfeld and Nicholson, 1967), 105–06. The best discussion about the belief in
white cannibalism among Africans is William Piersen, "White Cannibals, Black Martyrs: Fear,
Depression, and Religious Faith as causes of Suicide among New Slaves," *Journal of Negro
History* 62:2 (1977): 147–59.

10. Red wine was unknown for many African slaves carried to the New World. See Moreau
de Saint-Méry, *Description topographique,* 1:39; Antoine Simon Le Page du Pratz, *The History
of Louisiana,* ed. Joseph G. Tregle Jr. (1774; reprint, Baton Rouge: Louisiana State University Press, 1975), 357–58. Both Moreau de Saint-Méry and Le Page du Pratz are discussed in

Piersen, "White Cannibals, Black Martyrs," 149–50. These findings are confirmed by scholars who have studied how Africans often interpreted European demands for slaves as acts of vampirism or cannibalism. See Joseph C. Miller, *Way of Death: Merchant Capitalism and the Angolan Slave Trade* (Madison:University of Wisconsin Press, 1988), 4–5. For a more recent approach, see Luise White, *Speaking with Vampires: Rumor and History in Colonial Africa* (Berkeley: University of California Press, 2000).

11. Very often, the shark-mauled bodies of those who succeeded in drowning themselves were easily visible along the coastlines, not far from the European slave factories. See Richard B. Sheridan, "The Guinea Surgeons on the Middle Passage: The Provision of Medical Services in the British Slave Trade," *International Journal of African Historical Studies* 14:4 (1981): 601–25; Richard H. Steckel and Richard A. Jensen, "New Evidence on the Causes of Slave and Crew Mortality in the Atlantic Slave Trade," *Journal of Economic History* 46:1 (1986): 57–77; Elizabeth Donnan, ed., *Documents Illustrative of the Slave Trade to America* (New York: Hippocrene Books, 1965), 1:401–07.

12. Donnan, ed., *Documents*, 1:401–07.

13. Ibid.

14. Francisco Barrera y Domingo, *Reflexiones histórico físico naturales médico quirúrgicas. Prácticos y especulativos entretenimientos acerca de la vida, usos, costumbres, alimentos, bestidos, color y enfermedades a que propenden los negros de África, venidos a las Américas* (Havana: C.R., 1953), 68–74. In the logbook of the English ship James (1675–1676), two suicides were recorded while the ship was at sea. Both slaves—a woman and a man—were bought in Wyembah. She died after many days of "not eating or taking anything"; he "leaped over the boord [sic] and drowned himself." Donnan, ed., *Documents*, 1:401–07.

15. For instance, Moreau de Saint-Méry strongly recommended a *"grande surveillance"* with Igbo slaves, since they were often ready to kill themselves in the belief that they could then return to their homeland. Moreau de Saint-Méry, *Description topographique*, 1:51. In both Haiti and South Carolina, slaves from the Igbo and Calabar regions—the Niger Delta—were regarded as suicidal. Melville Herskovits, *The Myth of the Negro Past* (New York: Harper Brothers, 1941), 36–37. Michael Gomez has found some interesting testimonies given by African slaves in the antebellum South. Gomez discusses the so-called myth of the "flying Africans" mentioned by slaves from Georgia, South Carolina, and North Carolina. Gomez, *Exchanging Our Country Marks*, 116–19.

16. Spain and Great Britain signed two treaties to abolish the transatlantic slave trade. Both were virtually ignored by Spanish and Cuban slave merchants. See Murray, *Odious Commerce*, 72–113; Paquette, *Sugar Is Made with Blood*, 135, 139, 144–47.

17. José Antonio Bernal to Captain-General Vives. Havana, 11 August 1831. ANC: ME. 635/B.

18. The Carolinian Black Code was promulgated for the entire Spanish empire on 31 May 1789 under the influence of the Italian minister Jerónimo Grimaldi y Pallavicini, marquis of Grimaldi, from whom the document received its popular name, the Code of Grimaldi. See Rafael María de Labra, *Los códigos negros: Estudio de legislación comparada* (Madrid: Aurelio J. Aralia, 1879); Manuel Barcia, *Con el látigo de la ira*, 11–18, 85–94.

19. The term "Pythagorean beliefs" appears repeatedly in Cuban documents between 1788 and 1844. These "Phytagorean beliefs," based on the Greek philosopher's ideas about the transmigration of souls, was not unique to Cuba. Moreau de Saint-Méry used the same words while analyzing suicides among the Igbo slaves in Saint-Domingue. Moreau de Saint-Méry, *Description topographique*, 1:51.

20. Havana's merchants and planters to King Charles IV. ANC: RCJF. 150/7456.

21. Conclusion of public prosecutor Francisco Seidel. Havana, 8 September 1831. ANC: ME. 440/C.

22. Ibid.

23. Deposition of Vicente Pérez. Cafetal La Juanita, Mariel, December 1843. AHPPR: IJC. 1507/7754.

24. On 17 November 1844, José Cruz, a free black born in Africa and sentenced to death during La Escalera, hanged himself in prison. This created great uncertainty among the authorities in charge of with his condemnation. The prosecutor kept Cruz's body unburied until he received instructions about whether or not he should cut off Cruz's head. Cruz had, in fact, frustrated the whole ritual of demystification of the act of punishment arranged by the colonial courts. Since he had not died by a firing squad, prosecutor Gala was not sure whether beheading him would be useful in any way. Gala was sure that, as far as the rest of the African slaves were concerned, Cruz already had successfully gone back to his native land. Cruz's body was eventually buried without any mutilation. Apolinar de la Gala to Fulgencio Salas. Matanzas, 17 November 1844. ANC: ME. 38/1.

25. Claudio Martínez de Pinillos, count of Villanueva, to Captain-General Leopoldo O'Donnell. Havana, 30 June 1847. ANC: AP. 141/12.

26. Communication of Luis Hernandez. Los Acostas, Guane. March 1825. AHPPR: IJC. 1618/7953.

27. Deposition of Juan Benítez. June, 1844. AHPPR: IJC. 1507/7757.

28. Deposition of Vicente Pérez. Cafetal La Juanita, Mariel, December 1843. AHPPR: IJC. 1507/7754. Reis has noted how something similar occurred among the Nagô (that is, Lucumí slaves in Brazil) in Salvador da Bahia, writing, "Nagôs were the most militant. They were generally big people, well-built, robust, and willing to face death in their best clothes, believing that such a death is a way to return to their fatherland." Reis, *Slave Rebellion in Brazil*, 69.

29. Report on the suicide of Paulino Lucumí. Ingenio San Sebastián, El Mariel, June 1846. AHPPR: IJC. 1507/7761.

30. Although the ethnic denominations used in Cuba were highly arbitrary, plantation records and other documents refer to a majority of slaves with Congo, Carabalí, Mandinga, and Bambara ethnicities until the beginning of the nineteenth century. After the second decade of the century, Congo slaves still were the largest ethnic group in the island, but Lucumí slaves coming from the then-falling Empire of Oyo in Nigeria and its surroundings were the second-largest group in the western part of Cuba. By the mid-1850s Lucumís were a majority, with Carabalís, Congos, and Gangás in second, third, and fourth places respectively. See Moreno Fraginals, *El ingenio*, 2: 8–10.

31. The count of Villanueva to Captain-General O'Donnell. Havana, 30 June 1847. ANC: AP. 141/12.

32. Ibid.

33. Joaquín de Miranda y Madariaga to Captain-General Francisco Dionisio Vives. Cafetal plantation Reunión, 9 January 1827. ANC: RCJF. 150/7436.

34. Testimony of Domingo Lucumí. Santa Ana Jurisdiction, July 1835. ANC: ME. 232/Z.

35. In February 1843, the slave Vidal Ausas jumped into the steam-powered grinding machine of the sugar mill El Gato, owned by the marquis of Campo Florido and situated nearby the town of Güines, south of Havana. According to the Lieutenant José María Payá, the reasons for the suicide of this Hausa slave were not clear. José María Payá to the captain-general. Catalina, 12 February 1843. ANC: ME. 4282/X. Moreno Fraginals has commented about this type of suicide; see El ingenio, 2: chaps. 1 and 2.

36. Journal of José Pérez Sánchez. January 1829. ANC: RCJF. 150/7447.

37. José Pérez Sánchez to the prior and consuls of Havana. El Rubí, 31 December 1830. ANC: RCJF. 150/7456.

38. José Ramos to the governor of Matanzas. Sabanilla del Encomendador, 22 February 1832. AHPM: GP. 7/11; Trial against Nicolás Pentón due to the death of one slave and the suicide of another. El Algodonal, Banao, 1839. ANC: ME. 532/A.

39. Report of public prosecutor Manuel Martínez Serrano. Cayajabos, 1 October 1823. ANC: ME. 2433/G. Martínez Serrano recommended Nicolás's owner treat him with humanity due to his current state, both physical and mental.

40. One of the few recorded cases of bodily incineration took place just after the 1825 rebellion of Guamacaro was put down, when Alejandro Carabalí, from the coffee plantation La Hermita, was found hanging from a tree. Without hesitation and to set an example, Lieutenant Andrés Máximo Oliver ordered Alejandro's body burned immediately. Official communication of the Lieutenant Andrés Máximo Oliver. Guamacaro, 27 June 1825. ANC: CM. 1/3.

41. The count of Villanueva to Captain-General O'Donnell. Havana, 30 June 1847. ANC: AP. 141/12.

42. See Manuel Barcia, "El cementerio de los protestantes de la Habana," Boletín del Gabinete de Arqueología 1:1 (2001): 78–83.

43. The same can be said about infanticide practices. I have been able to find a few recorded cases in Cuban archives. Unfortunately the information contained in these files is poor and incomplete. See cases occurred in 1838 (ANC: ME. 1256/A); 1839 (ANC: ME. 1192/Y); and 1868 (ANC: ME. 1147/G); and also a failed attempt in 1828 (ANC: ME. 810/Ae). A last possible case occurred in 1841 (ANC: ME. 1227/J).

44. Information of the defender lieutenant Juan Gregorio Reyes. Havana, 1838. ANC: ME. 1044/Ak.

45. Juan Bernardo O'Gavan y Guerra, Observaciones sobre la suerte de los negros del Africa, considerados en su propia patria, y trasplantados á las Antillas españolas: y reclamación contra el tratado celebrado con los ingleses el año de 1817 (Madrid: Imprenta del Universal, 1821), 4–5.

46. Ibid., 9.

47. Juan Bernardo O'Gavan was born in 1783 in Santiago de Cuba, the second city of the island. In the late 1790s he went to Havana to begin his studies for the priesthood. During his eventful life, he was twice elected as a deputy to the Spanish courts, assumed care of the bishopric of Havana for some years, and enjoyed great influence among governors and colonial officials of Cuba.

48. The count of Villanueva to Captain-General O'Donnell. Havana, 30 June 1847. ANC: AP. 141/12.

49. The best proofs of this situation were the answers given by a group of planters who were asked by Captain-General Valdés about possible improvements to the situation of slaves on the plantations. Villanueva was the most influential Cuban-born character in the nineteenth century, with perhaps the sole exception of Francisco de Arango y Parreño. Villanueva was the man who represented the Council of Havana in Spain during the Napoleonic invasion of the peninsula. Later he was named treasurer of the island for some years until he became Intendente de Hacienda, a position that allowed him to manage, virtually alone, the Cuban treasury for more than twenty-five years. A Cuban-born functionary would not enjoy similar authority on the Cuban political and economic scene until the establishment of the Republic in 1902.

50. The marquis of Arcos to Captain-General Jerónimo Valdés. Havana, 19 May 1842. ANC: GSC. 941/33186.

51. The count of Fernandina to Captain-General Jerónimo Valdés. Havana, 12 March 1842. ANC: GSC. 940 / 33158.

52. Any scholar who investigates slave suicides in mid-nineteenth-century Cuba will face the dilemma of having a large amount of criminal records that refer to particular cases and a near-total absence of reliable statistics on actual extent of this phenomenon among slaves. In this situation, historians have been forced to work with what they have at their disposal, or, in Louis A. Perez's words, with statistics "mostly in fragmentary form." Perez, To Die in Cuba, 41.

53. Prosecutor Ignacio González Olivares to Captain-General O'Donnell. Havana, 6 December 1845. ANC: AP. 141/12.

54. The count of Villanueva to Captain-General O'Donnell. Havana, 30 June 1847. ANC: AP. 141/12.

55. See Émile Durkheim's classification of suicides in his Le suicide. Étude de sociologie (Paris: F. Alcan, 1897).

CHAPTER 5

1. Deposition of José María Lucumí. Nueva Filipina, July 1843. AHPPR: IJC. 35/76.

2. Depositions of Luciano Carabalí, Andrés Lalá, and Antonio José Carabalí. Nueva Filipina, July–August 1843. AHPPR: IJC. 35/76.

3. Conclusion of the criminal proceedings against Vicente Espinosa. Nueva Filipina, August 1843. AHPPR: IJC. 35/76.

4. The collections Asuntos Políticos, Comisión Militar, and Miscelánea de Expedientes in the Archivo Nacional de Cuba contain several cases of this type, some of which are examined throughout this chapter. For more examples, see Manuel Barcia, La resistencia esclava en las

plantaciones cubanas, 1790–1870 (Pinar del Río: Vitral, 1998), 14–19; Gloria García, *La esclavitud*, 91–168; María del Carmen Barcia, *La otra familia*, 47–80.

5. Cuban scholars only recently have begun to address systematically the use of the Spanish courts by eighteenth- and nineteenth-century slaves. Consequently, there are not quantitative studies on the subject, and existing statistics are scarce and mainly related to the late period of slavery in Cuba. This situation has been highlighted by Alejandro de la Fuente, who had to rely upon an isolated example to document the statistical analysis of his study on this topic. De la Fuente quoted a report by the Síndico Segundo of Havana in 1861, in which the syndic stated that 307 slaves had come to him asking for his intervention. Alejandro de la Fuente, "Su 'único derecho': Los esclavos y la ley," in *Debates y perspectivas: Cuadernos de historia y ciencias sociales* 4 (2004): 15 n. 13.

6. These collections of laws were the Ordenanzas de Santo Domingo of 1521 and the Ordenanzas de Cáceres of 1574.

7. An Act for the Better Ordering and Governing of Negroes (Barbados, 27 September 1661), NA: CO. 30/2/16–26, 25–28, and 32–33. Many of the slave codes referred in this work have been totally or partially reproduced in Stanley Engerman, Seymour Drescher, and Robert Paquette, eds., *Slavery* (Oxford: Oxford University Press, 2001), 92–148.

8. An Act for the Better Ordering and Governing of Negroes, clauses 2–7, 11, and 12.

9. Seventeenth-century British legislation was undergoing some important transformations, mostly due to religious discrimination. Some of these laws, especially the Clarendon Code (1661–1665) and the Test Act (1673) were particularly harsh. See Brian Magee, *The English Recusants* (London: Burns, Oats, and Washbourne, 1938) and E. I. Watkin, *Roman Catholicism in England from the Reformation to 1950* (London: Oxford University Press, 1957).

10. *Le Code Noir, ou recueil des règlements rendus jusqu'à present* (Paris: Prault, 1767).

11. A reproduction of the Danish slave code can be found in Aimery Caron and Arnold Highfield, *The French Intervention in the St. John Slave Revolt of 1733–34* (St. Thomas, V.I.: Bureau of Libraries, Museums, and Archaeological Services, Dept. of Conservation and Cultural Affairs, 1981), 15–16.

12. A reproduction of the Virginia slave code can be found in William Waller Hening, ed., *The Statutes at Large, Being a Collection of All the Laws of Virginia from the First Session of the Legislature in the Year 1619* (New York: R. & W. & G. Bartow, 1823), 3:447, 460–61. For New York, see Carl Nordstrom, "The New York Slave Code," *Afro-Americans in New York Life and History* 4:1 (1980): 7–26.

13. See *The Pictorial Times*, 17 June 1843, 212. See also the frequent references to mistreatments and abuses of slaves that appeared in the *British and Foreign Anti-Slavery Reporter* between 1840 and 1848.

14. Jorge Benci, *Economía cristã dos senhores no governo dos escravos*, ed. Serafim Leite, S.I. (1700; reprint, Oporto: Livraria Apostolado da Imprensa, 1954); Manoel Ribeiro Rocha, *Etíope resgatado, empenhado, sustentado, corrigido, instruído, e libertado*, ed. Paulo Suess (1758; reprint, Petrópolis: Editora Vozes, 1992).

15. In his attempt to systematize slave life, Rocha addressed several crucial issues related to the maintenance, punishment, religious instruction, and working duties of slaves, not

always from a very enlightened position. His ideas nonetheless influenced changes in the governance of slaves across Portuguese America, perhaps more than any of his predecessors. For an analysis of his ideas and propositions, see Marquese, Idéias, 16–17. See also Antonio Penalves Rocha, A economia política na sociedade escravista (São Paulo: HUCITEC, 1996), 142–45.

16. On the Brazilian case, see Marquese, Idéias. For the Rio Branco law of 1871, see Robert Conrad, The Destruction of Brazilian Slavery, 1850–1888 (Berkeley: University of California Press, 1973), 305–09.

17. See Elsa V. Goveia, The West Indian Slave Laws of the Eighteenth Century (Kingston: Caribbean University Press, 1970); Watson, Slave Law in the Americas; Mindie Lazarus Black and Susan F. Hirsch, eds., Contested States: Law, Hegemony, and Resistance (New York: Routledge, 1994); Timothy J. Paulson, Days of Sorrow, Years of Glory, 1813–1850: From the Nat Turner Revolt to the Fugitive Slave Law (New York: Chelsea House, 1994); Paul Finkelman, ed., Slavery and the Law (Madison: University of Wisconsin Press, 1997); Mark V. Tushnet, Slave Law in the American South: State v. Mann in History and Literature (Lawrence: University Press of Kansas, 2003).

18. Adele Hast, "The Legal Status of the Negro in Virginia, 1705–1765," Journal of Negro History 54:3 (1969): 220, 222. See also Arnold A. Sio, "Interpretations of Slavery: The Slave Status in the Americas," Comparative Studies in Society and History 7:3 (1965): 289–308; Christopher Morris, "The Articulation of Two Worlds: The Master-Slave Relationship Reconsidered," Journal of American History 85:3 (1998): 982–1007. On manumission, see also T. Stephen Whitman, The Price of Freedom: Slavery and Manumission in Baltimore and Early National Maryland (Lexington: University Press of Kentucky, 1997); Judith Kelleher Schafer, Becoming Free, Remaining Free: Manumission and Enslavement in New Orleans, 1846–1862 (Baton Rouge: Louisiana State University Press, 2003).

19. Sio, "Interpretations," 291.

20. Kenneth M. Stampp, The Peculiar Institution: Slavery in the Antebellum South (New York: Vintage, 1956); Tannenbaum, Slave and Citizen.

21. Sio, "Interpretations," 295. This opinion has been endorsed by other scholars. See Wilbert E. Moore, "Slave Laws and the Social Structure," Journal of Negro History 26:2 (1941): 171–202; Daniel J. Flanigan, "Criminal Procedure in Slave Trials in the Antebellum South," Journal of Southern History 40:4 (1974): 537–64; Christopher Waldrep, "Substituting Law for the Lash: Emancipation and Legal Formalism in a Mississippi County Court," Journal of American History 82:4 (1996): 1425–51. For the West Indies, see Arthur L. Stinchcombe, "Freedom and Oppression of Slaves in the Eighteenth-Century Caribbean," American Sociological Review 59:6 (1994): 911–29; Roger N. Buckley, "The Admission of Slave Testimony at British Military Courts in the West Indies, 1800–1809," in A Turbulent Time, ed. Gaspar and Geggus, 226–50. For Brazil, see Sidney Chalhoub, Visões da liberdade (São Paulo: Companhia das Letras, 1990); Chalhoub, "Slaves, Freedmen, and the Politics of Freedom in Brazil: The Experience of Blacks in the City of Rio de Janeiro," Slavery & Abolition 9–10 (1988–89): 64–84; Keila Grinberg, Liberata. A lei da ambiguidade. As ações de liberdade da Corte de Apelação do Rio de Janeiro no século XIX (Rio de Janeiro: Relume Dumara, 1994).

22. Stinchcombe, "Freedom and Oppression," 922–23.

23. Ibid.

24. Karasch, *Slave Life in Rio de Janeiro.* See especially the chapter on manumissions.

25. Sweet's study of eighteenth-century Rio de Janeiro stands as a perfect complement to Karasch's nineteenth-century analysis. James H. Sweet, "Manumission in Rio de Janeiro, 1749–54: An African Perspective," *Slavery & Abolition* 24:1 (2003): 54–70.

26. Ibid., 60.

27. Keila Grinberg, "La manumisión, el género, y la ley en el Brasil del siglo XIX: El proceso legal de Liberata por su libertad," in *Debate y Perspectivas: Cuadernos de historia y ciencias sociales* 4 (2004): 89–104.

28. See Stuart Schwartz, "The Manumission of Slaves in Colonial Brazil: Bahia, 1684–1745," *Hispanic American Historical Review* 54 (1974): 603–35; Karasch, *Slave Life in Rio de Janeiro*; Hebe Mattos, *Das cores do silêncio: os significados da liberdade no sudeste escravista: Brasil século XIX* (Rio de Janeiro: Nova Fronteira, 1998); Chalhoub, *Visões da Liberdade*; Sweet, "Manumission."

29. Grinberg, "La manumisión," 93.

30. Ibid., 97.

31. Ibid., 98.

32. Ibid., 99.

33. The Siete Partidas are generally regarded as the most important legal code issued during the Middle Ages. Despite being created for a completely different population, the Siete Partidas were used throughout Spanish America until well into the eighteenth century. See Robert I. Burns, ed. *Las Siete Partidas*, trans. Samuel Parsons Scott. 5 vols. (Philadelphia: University of Pennsylvania Press, 2001).

34. In 1789, when the code was enacted, slaves in Cuba were still victims of harsh measures. Branding new slaves was a common occurrence at least until the early 1780s, and according to the Bando of 1779, published by Captain-General Diego José Navarro, any slave found carrying a weapon was sentenced have his hand pierced by a nail and receive a hundred lashes for the first offense. Should he repeat the offense, he would receive another hundred lashes and his hand would be cut off. Leví Marrero, *Cuba: Economía y sociedad* (Madrid: Playor, [1981]) 13:207.

35. A previous slave code, issued in 1784, was never put into take effect. See "Código Negro Carolino (1784)," in Louis Sala-Molins, *L'Afrique aux Amériques. Le Code Noir espagnol* (Paris: Presses Universitaires de France, 1992).

36. Noncompliance with the dispositions of the Black Code of 1789 was so apparent that in 1804 the king was forced to issue a royal order (the Real Cédula of 22 April 1804) to enforce its observance. This royal order was also neglected by Cuban authorities and planters. Ortiz, *Los negros esclavos,* 336.

37. Labra, *Los códigos negros,* 24.

38. I have discussed the characteristics of the Carolinian Black Code elsewhere. For a detailed comparison between the 1789, 1842, and 1844 Cuban Black Codes, see Manuel Barcia, *Con el látigo de la ira,* 11–36.

39. See Lucena Salmoral, *Los códigos negros de la América Española.* For the black code enacted by the governor of Matanzas, Cecilio Ayllón, in 1825, see *Reglamento de policía rural de la jurisdicción del gobierno de Matanzas.* ANC: GSC. 1469/57999. For the Puerto Rican black code

of 1826, see Pedro Tomás de Córdoba, *Memorias geográficas, históricas, económicas, y estadísticas de la isla de Puerto Rico* (San Juan: Oficina del Gobierno, 1833), vol. 5. A reproduction of this Black Code can be found online at: http://freepages.genealogy.rootsweb.com/~poncepr/reglamento.html.

40. Some of these planters were educated in Europe and the United States, with a humanistic background that ultimately led ninety-three of them to petition for the effective abolition of the slave trade to Cuba. See "Importante exposición de los hacendados de Matanzas al Gobernador Capitán General, pidiendo la supresión de la trata," in José Antonio Saco, *Historia de la esclavitud de la raza africana en el Nuevo Mundo y en especial en los países Américo-Hispanos* (Havana: Cultural S.A., 1938) 4:154–74. Most of these planters were ambitious men, mainly peninsulares, who only had one thing in mind: profits.

41. Sebastián de Lasa to Captain-General Valdés. Havana, 5 March 1842. ANC: GSC. 940/33158.

42. The count of Fernandina to Captain-General Valdés. Havana, 12 March 1842. ANC: GSC. 940/33158.

43. Domingo Aldama to Captain-General Valdés. Ingenio Santa Rosa, Sabanilla del Encomendador, 18 March 1842. ANC: GSC. 940/33158.

44. José Manuel Carrillo to Captain-General Valdés. Havana, 3 March 1842. ANC: GSC. 940/33158.

45. José Muñoz Izaguirre to Captain-General Valdés. Havana, 7 March 1842. ANC: GSC. 940/33158.

46. Wenceslao de Villaurrutia to Captain-General Valdés. Havana, 25 March 1842. ANC: GSC. 940/33158.

47. Ibid.

48. The count of Santovenia to Captain-General Valdés. Havana, 22 April 1843. ANC: GSC. 940/33210.

49. Chapter 7 of the slave code of 1789. See Barcia, *Con el látigo de la ira*, 88–89.

50. Articles 29, 30, and 31 of the slave code of 1842. See Barcia, *Con el látigo de la ira*, 99–100.

51. Alejandro de la Fuente, "Slave Law and Claims-Making in Cuba: The Tannenbaum Debate Revisited," *Law and History Review* 22:2 (2004): 339–70.

52. See the slave code of Valdés, articles 34–37. Also see Barcia, *Con el látigo de la ira*, 100–01.

53. Scott, *Slave Emancipation*, 13. See also Hubert S. Aimes, "Coartación: A Spanish Institution for the Advancement of Slaves into Freedom," *Yale Review* 17 (1909): 412–31.

54. There are strong indications that slaves even owned horses. In 1842, for example, both Joaquín Muñoz Izaguirre and Juan Montalvo mentioned that their slaves sometimes owned mares with the intention of breeding them and selling the colts. Joaquín Muñoz Izaguirre to Captain-General Valdés. Havana, 7 March 1842; Juan Montalvo to Captain-General Valdés. Canasí, 8 April 1842. ANC: GSC. 940/33158. The rights to grow vegetables and to own small animals were sanctioned in both the slave code of 1789 and the slave code of Valdés in 1842. See Barcia, *Con el látigo de la ira*, 22–23.

55. The right to cultivate these land plots was legally sanctioned by chapter 3 of the slave code of 1789 and by article 13 of the slave code of 1842. See Barcia, *Con el látigo de la ira*, 87, 97.

56. Jacinto González Larrinaga to Captain-General Valdés. San Antonio de los Baños, 14 April 1842. ANC: GSC. 941/33186.

57. Domingo Aldama to Captain-General Valdés. Ingenio Santa Rosa, Sabanilla del Encomendador, 18 March 1842. ANC: GSC. 940/33158.

58. García, La esclavitud, 5

59. See Martínez Alier, Marriage, Class, and Colour in Nineteenth-Century Cuba; Maria del Carmen Barcia, La otra familia.

60. [José Agustín Caballero], "Instrucciones que ha dexado un Mayoral de azucarería a sus herederos," Papel Periódico de La Habana 50 (24 July 1791): 234–36.

61. Juan Montalvo to Valdés. Canasí, 8 April 1842. ANC: GSC. 940/33158.

62. See, for instance, Fick, The Making of Haiti, 91–97, 240–44, 260–63; Paquette, "The Drivers Shall Lead them," in Slavery, Secession, and Southern History, ed. Paquette and Ferleger, 31–58; Manuel Barcia, "Un aspecto de las relaciones de dominación," 88–93.

63. See the chapters on methods of slave punishment in García, La esclavitud, and Manuel Barcia, Con el látigo de la ira.

64. Anonymous letter received by Captain-General Francisco Serrano in July 1862. ANC: GSC. 954/33752.

65. Prosecutor resume and depositions of the slaves Eduardo, Diego, Bernardo, Pedro Congo, Mariana, Leonor, Hilaria, Isabel, Francisco, Antonio, and the Asiatic Achon, Miguel and Julio. ANC: ME. 1391/A.

66. The case against Esteban Santa Cruz de Oviedo seemed to have been dismissed. The file is inconclusive. ANC: GSC. 954/33752.

67. Prosecutor resume. ANC: ME. 1391/A.

68. See chapter 8 of the slave code of 1789 and article 41 of the slave code of 1842. Also see Manuel Barcia, Con el látigo de la ira, 89, 102.

69. Florencia Rodríguez to Captain-General Miguel Tacón. Havana, 20 November 1834. ANC: GSC. 936/33047.

70. The captain of Alacranes to the governor of Matanzas. Alacranes, 16 September 1839. AHPM: GP. 23/30.

71. The marquis of Cárdenas de Monte Hermoso to Captain-General Vives. Havana, 26 August 1828. ANC: ME. 440/C.

72. The count of Fernandina to García Oña. Matanzas, October 1843. ANC: ME. 4353/Bl.

73. Ibid.

74. Proceedings against six rebel slaves from the tobacco plantation Santa Mónica in San Diego de los Baños, which occurred on 15 June 1844. ANC: ME. 629/Af.

75. Certification written by José María de Herrera, count of Fernandina. Havana, 7 September 1844. ANC: ME. 629/Af. All the following details and quotations are from this document.

76. Ibid.

77. José Antonio Ramos claiming his right of manumission. Pinar del Río, July 1835. AHPPR: IJC. 29/65.

78. Criminal proceedings against Francisco Javier Congo for attempting suicide. Villa Clara, June 1838. ANC: ME. 1184/Aa.

79. Marcial Carabalí against his owner, Gregorio de Zayas. 1846–September 1852. AHPPR: IJC. 1506/7752.

80. Tomás Ortega to Cecilio Ayllón. Santa Ana, 4 August 1827. AHPM: GP. 23/13.

81. Félix Martínez to Captain-General Vives. Pipián, 28 May 1827. ANC: GSC. 936'/33024.

82. Deposition of Vicente Carabalí. Santiago de las Vegas, September 1835. ANC: ME. 568/M.

83. Deposition of Antonio Mandinga. Santiago de las Vegas, September 1835. ANC: ME. 568/M.

84. The sources for the study of slave resistance prior to 1800 are few, and many of the documents are in an appalling state of conservation. The fact that the slave population in Cuba begin its meteoric rise only after 1792 may also account for the paltry number of sources before then.

85. Depositions of the owner Paulino Iglesias, mayoral Esteban Mantilla, and contramayoral Antonio Carabalí. San Juan y Martínez, February 1826. AHPPR: IJC. 35/74.

86. Deposition of Manuel Iquiabo. San Juan y Martínez, February 1826. AHPPR: IJC. 35/74.

87. Deposition of Antonio Mandinga. San Juan y Martínez, February 1826. AHPPR: IJC. 35/74.

88. Proceedings against the mayoral Esteban Mantilla. San Juan y Martínez, February 1826. AHPPR: IJC. 35/74.

89. Report of prosecutor Blas Socarrás. 19 April 1828. AHPPR: IJC. 35/74.

90. Unfortunately, the final outcome of the trial is missing. Proceedings against Pedro Luis del Prado for the death of the slave Casimiro. San Diego, January 1840. AHPPR: IJC. 35/75.

91. Joseph T. Crawford to Lord Palmerston. Havana, 28 January 1848. NA: FO. 72/748.

CHAPTER 6

1. Report written by Manuel de Jesús Mata. Güira de Melena, October 1837. ANC: ME. 692/Q.

2. Deposition of Anastasio Lucumí. Güira de Melena, October 1837. ANC: ME. 692/Q. The other slaves were three Creoles, one Congo, and another Lucumí. The owner of the drums was Ignacio Mina, the cook on the estate of Benito Rodríguez.

3. Fernando Ortiz, Contrapunteo cubano del tabaco y el azúcar (Caracas: Biblioteca Ayacucho, 1947). Some recent studies on transculturation in the Americas are: Ivor L. Miller, "Religious Symbolism in Cuban Political Performance," Drama Review 44: 2 (2000): 30–55; Adrian Hearn, "Transformation, Transcendence, or Transculturation? The Many Faces of Cuban Santería," Humanities Research Centre 10: 1 (2003): 56–62; Misha Kokotovic, The Colonial Divide in Peruvian Narrative: Social Conflict and Transculturation (Brighton: Sussex Academic Press, 2005); Silvia Spitta, Between Two Waters: Narratives of Transculturation in Latin America (College Station: Texas A&M University Press, 2006).

4. However slippery that identifying intentionality may seem, as scholars we are entitled to have our perceptions of reality after spending time trying to understand the historical record. At times, identifying intention can be a treacherous enterprise. How to prove, for

example, that stealing a pig was an act of resistance? Answers to questions like this one are found within the context of the specific historical event. Who owned the pig? What did the slaves gain by stealing the pig? Were they taking revenge? Were they improving their living conditions by earning some extra income by selling the meat? By asking the proper questions, we may be able to gain a better understanding of the probable reasons behind the actions, although, of course, we will never be able to get into the heads of men and women who lived two centuries ago.

5. Carlos Ghersi to the captain-general of the island. 5 June 1843. ANC: GSC. 942/33246.

6. Proceedings against the slaves from the cafetal Nueva Empresa, owned by Antonio González Larrinaga, and the cafetal Asunción. Güira de Melena, September 1831. ANC: GSC. 936/33035. See also a brawl between Lucumí slaves in 1839 (ANC: ME. 2510/N); the problems between the slaves of the cafetal Solitario in 1834 (ANC: ME. 257/As); the fracas between the slaves of the ingenio Rosario in 1837 (ANC: ME. 199/Af); and other cases occurring in 1805 (ANC: ME. 808/J), 1832 (ANC: ME. 546/F), and 1835 between Carabalí slaves (ANC: ME. 1130/S).

7. Benito García y Santos to Captain-General O'Donnell. Macuriges, 12 May 1845. ANC: GSC. 942/33246.

8. In August 1834, some problems arose on the ingenio Santísima Trinidad in Güines. Two of the slaves involved were not on the plantation at the time. Lorenzo Gangá was in a neighboring tavern, while Domingo Mandinga was relaxing on another estate in the vicinity. Depositions of Lorenzo Gangá and Domingo Mandinga. Güines, August 1834. ANC: ME. 1134.

9. Deposition of Ignacio Zabaleta Criollo. Cayajabos, April 1844. AHPPR: IJC. 1509/7787. Ignacio was whipped by the ox-driver of the plantation for selling his plantains at the door of the infirmary without having a license and for taking sugarcane from the grinding mill on his way out. The ox-driver, curiously, was sentenced to spend seven months in prison for punishing the slave.

10. Deposition of Martín Carabalí. Guanabo, August 1831. ANC: ME. 490/F. Martín Carabalí was tried for beheading his concubine Felicia Arará, who was seeing a slave from a neighboring plantation. He was sentenced to death on 20 September 1831.

11. See Theresa Singleton, "Slavery and Spatial Dialectics on Cuban Coffee Plantations," World Archaeology 33:1 (2001): 98–114; Lisette Roura Alvarez, "Tercera campaña arqueológica en el antiguo cafetal El Padre," Gabinete de Arqueología: Boletín 2:2 (2002): 133.

12. In the collection "Miscelánea de Expedientes" of the Archivo Nacional de Cuba, there are several cases of fires occurring on sugar and coffee plantations in the first half of the nineteenth century. See ANC: ME, bundles 804, 805, and 807. This topic deserves serious study due to its impact on several aspects of Cuban society. Reynaldo Funes's study about the deforestation of Cuba is perhaps the closest we currently have on the topic. See Reinaldo Funes Monzote, De Bosque a sabana: Azúcar, deforestación, y medio ambiente en Cuba, 1492–1926 (Mexico D.F.: Siglo XXI, 2004).

13. Fires were set by slaves, for example, in the revolt of 1825 in Guamacaro, the revolt of 1833 in Guanajay, and on the sugar mills Santa Rosa and La Majagua in May 1843. At other times, slaves' attempts to set plantations houses ablaze failed, as happened at the sugar mill La Arratía in July 1842.

14. For example, in July 1837, the slave Juan Quevedo in the jurisdiction of Cimarrones planned to set fire to the newly installed grinding steam machine because, as Luis Gangá explained, he "did not want to grind the sugar cane in the steam machine that their owner was installing." Deposition of Luis Gangá. Cimarrones, July 1837. ANC: ME. 609/R.

15. Deposition of Eduardo Hernández, natural from New Orleans, and co-owner and administrator of the sugar mill Unión. March 1848. ANC: ME. 1207/A.

16. There were more fires during 1844, but this one was seen as a statement. For instance, in February, the sugar mills Botino, Perla, and Paz, all located in the jurisdiction of Guamacaro, caught fire simultaneously. Cayetano Peñalver to the governor of Matanzas, Antonio García Oña. Guamacaro, 11 February 1844. AHPM: GP. 8/11.

17. "Incendio en la Santísima Trinidad de Oviedo," Sabanilla del Encomendador, 13 May 1844. AHPM: GP. 9/38.

18. Deposition of Francisco Casañas. Camarioca, February 1839. ANC: GSC. 939/33127. According to Robert L. Paquette and Joseph C. Dorsey, for several years these slaves had "defied, insulted, and threatened whites. They had brutalized a mayoral, set fires, robbed neighboring estates, and plotted an insurrection." Paquette and Dorsey, "The Escoto Papers and Cuban Slave Resistance," *Slavery & Abolition* 15:3 (1994): 90–91.

19. Curiously, the ingenio Jesús María was administered by Esteban Santa Cruz de Oviedo. Trinidad was finally captured. After being arrested he tried to commit suicide by stabbing himself in the stomach. He survived. Joaquín Ramos to the governor of Matanzas. Sabanilla del Encomendador, 22 February 1832. AHPM: GP. 7/11.

20. Deposition of owner José del Calvo. April 1846. AHPPR: IJC. 1507/7760.

21. Deposition of Francisco de Paula Ramos. April 1846. AHPPR: IJC. 1507/7760.

22. Deposition of Joaquín Curbelo. Partido de Catalina, September 1837. ANC: ME. 1193/H.

23. Rafael O'Farrill to Captain-General Valdés. Havana, 26 February 1842. ANC: GSC. 940/33158.

24. Deposition of Cristóbal Carabalí. Cafetal Santa Catalina, Guanajay, July 1828. ANC: GSC. 936/33025. Cristóbal was also an old offender maroon and he identified himself as such to the authorities. He also confessed to spending most of his time in the hospital of the plantation because he was always ill.

25. Deposition of Bárbara Gangá. Ingenio San Matías, Aguacate, November 1829. ANC: ME. 1246/C.

26. Sentence against José del Rosario Criollo for murdering of his contramayoral, Francisco (alias Chico) Criollo. Aguacate, November 1829. ANC: ME. 1246/C.

27. Proceedings against Salvador de la Paz Martiartu. September 1843. ANC: GSC. 941/33159. His other plantation, the ingenio La Arratía, suffered two slave revolts between 1842 and 1843.

28. Deposition of Pedro Cano. Matanzas, September 1843. ANC: GSC. 941/33159.

29. Wenceslao de Villaurrutia to Valdés. Havana, 25 March 1842. ANC: GSC. 940/33158.

30. Ibid. Villaurrutia was referring to the revolt of 1825 in Guamacaro. Despite the assassination of various slave owners in the region and the fact that the uprising was not organized

or led by Fouquier's slaves, the revolt was remembered in the following decades as the "revolt of Fouquier's slaves."

31. There were several conspiracies and revolts staged coffee plantations throughout the first half of the nineteenth century. See chapter 2.`

32. Collective protests were also common in other places of the New World. See, for example, Stuart Schwartz, "Resistance and Accommodation in Eighteenth-Century Brazil: The Slaves' View of Slavery," *Hispanic American Historical Review* 57 (1977): 69–81; João José Reis, "The Revolution of the Ganhadores: Urban Labor, Ethnicity, and the African Strike of 1857 in Bahia, Brazil," *Journal of Latin American Studies* 29:2 (1997): 355–93.

33. The marquis of Aguas Claras to Manuel González, Captain of the jurisdiction of Tapaste. Cafetal La Asunción, 12 August 1843. ANC: ME. 4359/Bd.

34. Deposition of Feliciano Carabalí. Cafetal Santa Catalina, July 1828. ANC: GSC. 936/33025.

35. Francisco Fernández de Castro to Captain-General Valdés. Bejucal, 26 May 1842. ANC: GSC. 941/33196.

36. Proceedings against the disobedient slaves from the ingenio Nueva Vizcaya (a) Camarones, owned by Santiago Garmony. Yumurí, 1837. ANC: ME. 616/Ad.

37. Criminal proceedings against the rebel slaves in the ingenio Santísima Trinidad. Güines, 15 August 1834. ANC: ME. 1134/M.

38. See Chapter 2.

39. A small sample of these cases may be found in ANC: ME. 237/I; 277/J; 415/F; 430/A; 545/F; 794/E; and 1069/B.

40. Criminal proceedings against Antonio Viví, Elías Gangá, and Bernabé Congo for abducting and murdering an ox. Guanabo, 1836. ANC: ME. 459/V.

41. Deposition of José Manresa in the proceedings against Francisco Gangá for assault. Aguacate, October 1836. ANC: ME. 684/All.

42. Deposition of Francisco Díaz. Guanajay, March 1837; Pedro Soulé de Limendoux to the president of the Military Commission. Havana, 5 June 1837. ANC: ME. 277/J.

43. Deposition of Manuel Cabrera in the proceedings against Juan Bautista, José Luis, and José Nicolás, slaves of Manuel Mena, for assault. ANC: ME. 430/A

44. Throughout the nineteenth century, capital punishment was the usual penalty for slaves who murdered other slaves. If a slave killed a white person, the outcome was almost certainly the death penalty.

45. Deposition of Miguel Gangá. La Guanábana, November 1845. ANC: ME. 442/Ab.

46. Sentence of the proceedings against the four thieves who assaulted the potrero of the ingenio Santa Lutgarda. December 1845. ANC: ME. 442/Ab.

47. Deposition of Claudio Carabalí. Guanabacoa, 13 August 1834. ANC: ME. 256/Q.

48. Proceedings to inquire about the thieves who assaulted the ingenio Jesús María and killed one of his slaves. Guanabacoa, August 1834. ANC: ME. 256/Q.

49. See the circular letter issued by Captain-General Jerónimo Valdés in April 1842 prohibiting the owners of slaves to teach them how to use firearms, "since this brings as a result

that [the slaves] learn to use the firearms with grave prejudice to the tranquillity and security of the inhabitants of this island." Circular of Valdés, April 1844. ANC: GSC. 941/33159.

50. See, for example, Abiel Abbot, *Letters Written in the Interior of Cuba* (Boston: Bowles and Dearborn, 1829); Merlín, *Viaje a la Habana*; Hyppolite Piron, *La isla de Cuba* (Santiago de Cuba: Editorial Oriente, 1995).

51. The head had been placed at a crossroad as a grim reminder to other mulattos and blacks of the consequences that their actions could have. ANC: CM. 81/7.

52. If we accept the idea that a process of transculturation by which Judeo-Christian and African deities were merged took place in nineteenth-century Cuba, then we also should consider the possibility that attending mass or visiting church at any other time to worship those African deities disguised in European clothes could be—at the very least in exceptional cases—acts of resistance.

53. See, for instance, the censuses of 1774, 1792, 1817, and 1827 in Ramón de la Sagra, *Historia económica-política y estadística de la Isla de Cuba* (Havana: Imprenta de las viudas de Arazoza y Soler, 1831).

54. Conclusions of Military Commission prosecutor Francisco Seidel. May 1828. ANC: CM. 2/4.

55. Slave barracks were first ordered by the slave code for the province of Matanzas, issued on 22 October 1825 by Governor Cecilio Ayllón. See *Reglamento de Policía Rural de la Jurisdicción del Gobierno de Matanzas*, art. 14.

56. See Juan Pérez de la Riva, *El barracón y otros ensayos* (Havana: Ciencias Sociales, 1975); Scott, *Slave Emancipation in Cuba*, 17–19.

57. Deposition of Matías Gangá. Ceja de Pablo, March 1844. ANC: CM. 35/3.

58. Joaquín Muñoz Izaguirre to Captain-General Valdés. Havana, 7 March 1842. ANC: GSC. 940/33158.

59. José Manuel Carrillo to Captain-General Valdés. Havana, 3 March 1842. ANC: GSC. 940/33158.

60. José Leopoldo Yarini, untitled manuscript on the 1833 epidemic of cholera in his ingenio in Guamacaro. ACEHOC. Unclassified.

61. See more about these two slave uprisings in Chapter 2.

62. Deposition of José Ignacio Padilla. Hacienda La Llanada, Las Jaironas. July 1845. ANC: ME. 1148/Q.

63. Padilla also remembered how Alejo shouted diverse insults and blasphemies against God. Padilla was finally found guilty and condemned to pay a fine of twenty-five pesos and trial expenses. ANC: ME. 1148/Q.

64. Scott, *Slave Emancipation in Cuba*, 169–70. The letter was written by Vice-Consul Harris to Consul-General Crowe, 2 April 1883.

65. Groans, sights, moans, chuckles, well-timed silences, winks, or stares could all be forms of "thinly veiled dissent." Scott, *Domination and the Arts of Resistance*, 155.

66. According to Fernando Ortiz, "There are warrior dances where the dancers with their weapons and shields re-enact war episodes or predict those of victories to come." Ortiz, *Los bailes y el teatro de los negros en el folklore de Cuba*, 272.

67. Silence could also be a form of resistance. Yarini's slaves' collective refusal to sing, for example, could be easily understood as a form of resistance. African slaves and their descendants brought with them different types of drums. Some of these drums were used in both sacred and profane celebrations; others—such as the Carabalí Ekué and the Lucumí Batá—were only meant to be used in sacred events. See Ortiz, *Los instrumentos de la música afrocubana* and *Ensayos Etnográficos* (Havana: Ciencias Sociales, 1984). For an analysis of the same issues in Brazil, see João José Reis, "Batuque: African Drumming and Dance between Repression and Concession, Bahia, 1808–1855," *Bulletin of Latin American Studies* 24:2 (2005): 201–14.

68. The lyrics sung by the rebels of 1833 are reproduced in Chapter 2.

69. Stuart Schwartz, "Cantos and Quilombos: A Hausa Rebellion in Bahia, 1814," in *Slaves, Subjects, and Subversives*, ed. Landers and Robinson, 263.

70. Nicolás de Cárdenas to Carlos Ghersi. Ingenio La Conchita, 17 May 1839. ANC: CM. 133/2.

71. Deposition of Cleto Lucumí. Macuriges, May 1839. ANC: CM. 133/2.

72. Deposition of Marcelino Lucumí de Coto. Ingenio La Andrea, January 1844. ANC: CM. 37/1.

73. Tomás Betancourt y Ferrer to Francisco Yllás. Partido de Juan Angola, 4 May 1827. ANC: AP. 125/36.

74. José María Gavilán to Captain-General Joaquín de Ezpeleta. San Nicolás, 26 May 1838. ANC: GSC. 938/33102.

75. Fernando Ortiz, *Glosario de Afronegrismos* (Havana: Ciencias Sociales, 1990); Lydia Cabrera, *Refranes de negros viejos* (Miami: Ediciones C.R., 1970); Samuel Feijóo, *El negro en la literature folklorica cubana* (Havana: Letras Cubanas, 1980). All the sayings included here have been taken from these three books.

76. The totí or Cuban blackbird (*Dives atroviolacea*) is a Cuban black bird with a metallic sheen that can be found across the island. In Cuban culture, tradition has transformed the totí into a scapegoat par excellence.

77. This gathering seems to have been about improving the crops. The original transcription of this line is "buniatal, que no falta boniato, platanal que no falta platanos." Criminal proceedings carried out by Domingo Orúe, captain and judge of the jurisdiction of San Diego de los Baños. January 1845. ANC: ME. 1148/L.

78. Ibid.

79. Information written by the Lieutenant Luis de León. Aguacate, March 1839. ANC: ME. 448/Q.

80. Ibid.

81. Ibid.

82. Deposition of José Carabalí Isuamo. Aguacate, March 1839. ANC: ME. 448/Q.

83. Ibid.

84. Prosecutor Apolinar de la Gala's conclusions. March 1839. ANC: ME. 448/Q.

85. Deposition of Vicente Rodríguez, mayoral of the ingenio Santa Ana de Jaspe in Ceja de Pablo. April 1844. ANC: CM. 35/5.

86. Yarini, untitled manuscript.

87. Yarini, untitled manuscript.

88. See Chapter 2.

89. Yarini, untitled manuscript.

90. Scott, *Domination and the Arts of Resistance*, 28–29.

91. See the firsthand opinion of the nineteenth-century Swedish traveler Fredrika Bremer. Bremer, *The Homes of the New World: Impressions of America* (New York: Harper & Brothers, 1853), especially letters 34 and 35.

92. Eugene Genovese, *Roll, Jordan, Roll: The World the Slaves Made* (New York: Pantheon, 1974), 597–98. Also see Stephanie M. H. Camp, *Closer to Freedom: Enslaved Women and Everyday Resistance in the Plantation South* (Chapel Hill: University of North Carolina Press, 2004); Walker, *No More, No More.*

CONCLUSION

1. "Representación de la Ciudad de la Habana á las Cortes, el 20 de julio de 1811, con motivo de las proposiciones hechas por D. José Miguel Guridi Alcocer y D. Agustín de Argüelles, sobre el tráfico y esclavitud de los negros; extendida por el Alférez Mayor de la Ciudad, D. Francisco de Arango, por encargo del Ayuntamiento, Consulado y Sociedad Patriótica de la Habana." Francisco de Arango y Parreño, *Obras,* 2:175–76. Although the text was signed by a large number of important citizens of Havana, the real and only author of this document was Arango y Parreño. See also Alejandro E. Gómez, "El síndrome de Saint-Domingue: Percepciones y sensibilidades de la Revolución Haitiana en el Gran Caribe (1791–1814)," *Caravelle* 86 (2006): 125–55.

2. Fick, *The Making of Haiti*, 238.

3. See, for example, the opinion of Wenceslao de Villaurrutia in 1842. Villaurrutia to Captain-General Valdés. Havana, 25 March 1842. ANC: GSC. 940/33158.

4. Genovese, *Roll, Jordan, Roll*, 597.

5. Ibid., 598.

6. Reis and Silva, *Negociação e conflito*, 15.

7. Luis A. Figueroa, *Sugar, Slavery, and Freedom in Nineteenth-Century Puerto Rico* (Chapel Hill: University of North Carolina Press, 2005), 82.

8. Scott, *Domination and the Arts of Resistance*, 175.

9. Genovese, *From Rebellion to Revolution*, 6.

10. For a comparison of these two sequences of slave revolts, see Barcia, "Slave Rebellions."

11. Genovese, *Roll, Jordan, Roll*, 598.

12. Joaquín Muñoz Izaguirre to Captain-General Valdés. Havana, 7 March 1842. ANC: GSC. 940/33158.

13. On this issue, see Chapter 5.

14. Domingo Aldama to Captain-General Valdés. Ingenio Santa Rosa, Sabanilla del Encomendador. 18 March 1842. ANC: GSC. 940/33158.

15. Manuel Moreno Fraginals, "Africa in Cuba: A Quantitative Analysis of the African Populations in the Island of Cuba," in *Comparative Perspectives on Slavery in New World Plantation Societies,* ed. Vera Rubin and Arthur Tuden (New York: New York Academy of Sciences, 1977), 192.

Bibliography

ARCHIVAL SOURCES

Archivo Nacional de Cuba (ANC)
Asuntos Políticos (AP)
Comisión Militar (CM)
Gobierno Superior Civil (GSC)
Miscelánea de Expedientes (ME)
Real Consulado y Junta de Fomento

Archivo Histórico Provincial de Matanzas (AHPM)
Gobierno Provincial (GP)

Archivo Histórico Provincial de Pinar del Río (AHPPR)
Instituciones Judiciales Coloniales (IJC)

Archivo Histórico Provincial de Santiago de Cuba (AHPSC)
Gobierno Provincial (GP)

Archivo del Museo de Historia de la Ciencia "Carlos J. Finlay"

Archivo de la Oficina del Historiador de la Ciudad (Habana) (AOHCH)
Actas de Cabildo Trasuntadas (ACT)
Fondo General (FG)

National Archives (London) (NA)
Colonial Office (CO)
Foreign Office (FO)

NEWSPAPERS

Bohemia (Havana)
British and Foreign Anti-Slavery Reporter (London)
Diario del Gobierno de la Habana (Havana)
Hojas Literarias (Havana)
Papel Periódico de la Habana (Havana)
Revista Cubana (Havana)
The Times (London)
The Pictorial Times (London)

PRINTED PRIMARY SOURCES

Abbot, Abiel. *Letters Written in the Interior of Cuba.* Boston: Bowles and Dearborn, 1829.
Antonil, André João. *Cultura e opulência do Brasil por suas drogas e minas.* Trans. Andrée Mansuy. 1711; Paris: Institut des Hautes Études de l'Amérique Latine, 1968.
Bacon, Francis. *Works.* 3 vols. Ed. Basil Montagu. Philadelphia: Carey and Hart, 1844.
Barnet, Miguel, ed. *The Autobiography of a Runaway Slave: Esteban Montejo.* Trans. Jocasta Innes. New York: Pantheon, 1968.
Barrera y Domingo, Francisco. *Reflexiones histórico físico naturales médico quirúrgicas. Prácticos y especulativos entretenimientos acerca de la vida, usos, costumbres, alimentos, bestidos, color, y enfermedades a que propenden los negros de África, venidos a las Américas.* Havana: C.R., 1953.
Beccaria, Cesare, marchese di. *On Crimes and Punishments, and Other Writings.* Ed. Richard Bellamy. Trans. Richard Davies with Virginia Cox and Richard Bellamy. Cambridge: Cambridge University Press, 1995.
Benci, Jorge. *Economía cristã dos senhores no governo dos escravos.* Ed. Serafim Leite, S.I. 1700. Reprint, Oporto: Livraria Apostolado da Imprensa, 1954.
Bosman, Willem. *A New and Accurate Description of the Coast of Guinea.* London: Frank Cass, 1705.
Bremer, Fredrika. *The Homes of the New World: Impressions of America.* Trans. Mary Howitt. New York: Harper & Brothers, 1853.
Burns Robert I., ed. *Las Siete Partidas.* 5 vols. Trans. Samuel Parsons Scott. Philadelphia: University of Pennsylvania Press, 2001.
Cantero, Justo Germán. *Los ingenios: Colección de vistas de los principales ingenios de azúcar de la isla de Cuba.* Havana: L. Marquier, 1857.
Córdoba, Pedro Tomás de. *Memorias geográficas, históricas, económicas, y estadísticas de la isla de Puerto Rico.* Vol. 5. San Juan: Oficina del Gobierno, 1833.
Cuadro estadístico de la siempre fiel Isla de Cuba, correspondiente al año 1827. Havana: Imprenta del Gobierno, 1818.
Dallas, Robert Charles. *The History of the Maroons, from Their Origin to the Establishment of Their Chief Tribe at Sierra Leone: Including the Expedition to Cuba, for the Purpose of*

Procuring Spanish Chasseurs and the State of the Island of Jamaica for the Last Ten Years, with a Succinct History of the Island Previous to That Period. 2 vols. London: T. N. Longman and O. Rees, 1803.

Donnan, Elizabeth, ed. Documents Illustrative of the Slave Trade to America, 4 vols. New York: Hippocrene Books, 1965.

Escudero, José Eusebio. Reflexiones que sobre los palenques de negros cimarrones respecto a la parte oriental de esta Ysla de Cuba, se elevaron al conocimiento de S. M. Santiago de Cuba, 1819.

[Estévez, Francisco.] Diario del rancheador. Ed. Cirilo Villaverde. Havana: Letras Cubanas, 1982.

Hening, William Waller, ed. The Statutes at Large, Being a Collection of All the Laws of Virginia from the First Session of the Legislature in the Year 1619. Vol. 3. New York: R. & W. & G. Bartow, 1823.

Holbach, Paul-Henri Thiry, baron d'. Système de la nature, ou, Des loix du monde physique & du monde moral. London: [n.p.], 1771.

Labra, Rafael María de. Los códigos negros. Estudio de legislación comparada. Madrid: Aurelio J. Aralia, 1879.

Le Code Noir, ou Recueil des reglements rendus jusqu'à présent. Paris: Prault, 1767.

Le Page du Pratz, Antoine Simon. The History of Louisiana. Ed. Joseph G. Tregle Jr. 1774. Reprint, Baton Rouge: Louisiana State University Press, 1975.

Merlin, María de las Mercedes Santa Cruz y Montalvo, condesa de. Viaje a la Habana. Madrid: Sociedad Literaria, 1841.

Montejo, Esteban. The Autobiography of a Runaway Slave. Ed. Miguel Barnet. Trans. Jocasta Innes. London: Bodley Head, 1966.

Montesquieu, Charles de Secondat, baron de. Lettres persanes. Ed. Paul Vernière. Paris: Garnier Frères, 1960.

Moreau de Saint-Méry, M. L. E. Description topographique, physique, civile, politique, et historique de la partie française de l'isle Saint-Domingue. 2 vols. Philadelphia: Chez l'Auteur, 1797.

Nuevo reglamento y arancel que debe gobernar en la captura de esclavos cimarrones, aprobado por S. M. en Real Orden expedida en San Lorenzo con fecha de veinte de diciembre de 1796. Havana: Imprenta de la Capitanía General, 1796.

O'Gavan y Guerra, Juan Bernardo. Observaciones sobre la suerte de los negros del Africa, considerados en su propia patria, y trasplantados á las Antillas españolas: y reclamación contra el tratado celebrado con los ingleses el año de 1817. Madrid: Imprenta del Universal, 1821.

Ordenações e leis de reino de Portugal, recopiladas por mandado dí El Rey D. Philippe I. 14th ed. Rio de Janeiro, 1870.

Paseo pintoresco por la isla de Cuba. Vol. 2. Havana: Impresa de Soler, 1842.

Piron, Hippolyte. La isla de Cuba. Ed. Olga Portuondo Zúñiga. Trans. Walfrido Díaz. Santiago de Cuba: Editorial Oriente, 1995.

Reglamento de policía rural de la jurisdicción del Gobierno de Matanzas. Matanzas: Imprenta del Gobierno de Matanzas, 1825.

Resumen del censo de población de la isla de Cuba a fin del año de 1841, formado de orden del Excmo. Sr. Capitán General de la misma D. Jerónimo Valdés. Havana: Imprenta de la Capitanía General, 1842.

Rocha, Manoel Ribeiro. *Etíope resgatado, empenhado, sustentado, corrigido, instruído, e libertado.* Ed. Paulo Suess. 1758. Reprint, Petrópolis: Editora Vozes, 1992.

Rugendas, J. P. M. *Voyage pittoresque dans le Brésil.* Paris: Engelmann & Cie, 1835.

Saco, José Antonio. *Historia de la esclavitud de la raza africana en el Nuevo Mundo y en especial en los países Américo-Hispanos.* 6 vols. 1879. Reprint, Havana: Cultural S.A., 1938.

Sagra, Ramón de la. *Historia económica-política y estadística de la Isla de Cuba.* Havana: Imprenta de las viudas de Arazoza y Soler, 1831.

Savary, Jacques. *Le parfait négociant.* 1675. Reprint, Lyon: Jacques Lyons, 1697.

Speke, John Hanning. *Journey of the Discovery of the Source of the Nile.* 1863. Reprint, New York: Greenwood Press, 1969.

SECONDARY SOURCES

Achille-Delmas, F. *Psychologie pathologique du suicide.* Paris: Librairie Félix Alcan, 1932.

Aguirre Beltrán, G. "The Rivers of Guinea." *Journal of Negro History* 31:3 (1946): 290–316.

Aimes, Hubert S. "Coartación: A Spanish Institution for the Advancement of Slaves into Freedom." *Yale Review* 17 (1909): 412–31.

Aptheker, Herbert. *American Negro Slave Revolts.* New York: Columbia University Press, 1943.

Arango y Parreño, Francisco de. *De la factoría a la colonia.* Havana: Publicaciones de la Secretaría de Educación La Habana, 1936.

Aremu, P. S. O. "Between Myth and Reality: Yoruba Egungun Costumes as Commemorative Clothes." *Journal of Black Studies* 22:1 (1991): 6–14.

Ashcroft, Bill, Gareth Griffiths, and Helen Tiffin. *Key Concepts in Post-Colonial Studies.* London: Routledge, 1998.

Augier, Angel. "José Antonio Aponte y la conspiración de 1812." *Bohemia* 54:15 (1962): 48–64.

Balbuena Gutiérrez, Bárbara. *El íreme Abakuá.* Havana: Pueblo y Educación, 1996.

Barcia, Manuel. *La resistencia esclava en las plantaciones cubanas, 1790–1870.* Pinar del Río: Vitral, 1998.

———. *Con el látigo de la ira: Legislación, represión, y control en las plantaciones cubanas.* Havana: Ciencias Sociales, 2000.

———. "La sublevacion de esclavos de 1825 en Guamacaro." M.A. thesis, University of Havana, 2000.

———. "El cementerio de los protestantes de la Habana." *Boletín del Gabinete de Arqueología* 1:1 (2001): 78–83.

———. "Un aspecto de las relaciones de dominación en la plantación esclavista cubana: Los contramayorales esclavos." *Boletín del Gabinete de Arqueología* 1:2 (2001): 88–93.

———. "Slave Rebellions in Latin America during the 'Age of Revolution': Bahia and Havana-Matanzas from a Comparative Perspective." M.A. thesis, University of Essex, 2002.

Barcia, María del Carmen. *Burguesía esclavista y abolición.* Havana: Ciencias Sociales, 1987.

———. *La otra familia: Parientes, redes, y descendencia de los esclavos en Cuba.* Havana: Casa de las Américas, 2003.

———, and Manuel Barcia. "La conspiración de La Escalera: El precio de una traición." *Catauro* 2:3 (2001): 199–204.

Barker, David, and Balfour Spence. "Afro-Caribbean Agriculture: A Jamaican Maroon Community in Transition." *Geographical Journal* 154:2 (July 1988): 198–208.

Basso Ortiz, Alessandra. "Los gangá longobá: El nacimiento de los dioses." *Boletín Antropológico* (Año 20) 2:52 (2001): 195–208.

———. "La rumba, ¿Género de origen Gangá?" *La Jiribilla* 65 (2002). Online at: www. lajiribilla.cu/2002/n65_agosto.

Bauer, Raymond A., and Alice H. Bauer. "Day-to-Day Resistance to Slavery." *Journal of Negro History* 27 (1942): 388–419.

Beauchamp, Tom. "An Analysis of Hume's Essay on Suicide." *Review of Metaphysics* 30 (1976): 73–95.

Beier, H. U. "Spirit Children among the Yoruba," *African Affairs* 53:213 (1954): 328–31.

Benneh, G. "Systems of Agriculture in Tropical Africa." *Economical Geography* 48 (1972): 244–57.

Bergad, Laird T. *Cuban Rural Society in the Nineteenth Century: The Social and Economic History of Monoculture in Matanzas.* Princeton: Princeton University Press, 1990.

———, Fe Iglesias García, and María del Carmen Barcia. *The Cuban Slave Market, 1790–1880.* Cambridge: Cambridge University Press, 1995.

Bilby, Kenneth. "Swearing by the Past, Swearing to the Future: Sacred Oaths, Alliances, and Treaties among the Guianese and Jamaican Maroons." *Ethnohistory* 44:4 (1997): 655–89.

Black, Mindie Lazarus, and Susan F. Hirsch, eds. *Contested States: Law, Hegemony, and Resistance.* New York: Routledge, 1994.

Blackburn, Robin. *The Overthrow of Colonial Slavery, 1776–1848.* London: Verso, 1988.

———. *The Making of New World Slavery: From the Baroque to the Modern, 1492–1800.* London: Verso, 1997.

Blier, Suzanne Preston. "The Path of the Leopard: Motherhood and Majesty in Early Danhome." *Journal of African History* 36:3 (1995): 391–417.

Body, Janice. "Spirit Possession Revisited: Beyond Instrumentality." *Annual Review of Anthropology* 23 (1994): 407–34.

Bolívar, Natalia. "Los Changaní de Guanabacoa: Reglas de Palo." Catauro 2:3 (2001): 211–19.

Bourdieu, Pierre. Outlines of a Theory of Practice. Trans. Richard Nice. Cambridge: Cambridge University Press, 1977.

Brana-Shute, Rosemary. "Approaching Freedom: The Manumission of Slaves in Suriname." Slavery & Abolition 10:3 (1989): 40–63.

Broadhead, Susan Herlin. "Beyond Decline: The Kingdom of the Kongo in the Eighteenth and Nineteenth Centuries." International Journal of African Historical Studies 12:4 (1979): 615–50.

Bronfman, Alejandra. "Reading Maceo's Skull (Or the Paradoxes of Race in Cuba)." Princeton University, Program in Latin American Studies Bulletin (Fall 1998): 17–18.

Brown, Carolyn. We Were All Slaves: African Miners, Culture, and Resistance at the Enugu Government Colliery, Nigeria. Oxford: James Currey, 2003.

Buckley, Roger N. "The Admission of Slave Testimony at British Military Courts in the West Indies, 1800–1809." In A Turbulent Time, ed. Gaspar and Geggus. 226–50.

Burke, Peter. History and Social Theory. Cambridge: Polity Press, 1992.

Cabrera, Lydia. Cuentos negros de Cuba. Havana: Ediciones Nuevo Mundo, 1961.

———. El Monte: Igbo, finda, ewe orisha, vititinfinda. Miami: Rema Press, 1968.

———. Refranes de negros viejos. Miami: Ediciones C.R., 1970.

———. Koeko Iyawo: Aprende Novicia. Pequeño tratado de regla Lucumí. Miami: Ultra Graphics, 1980.

———. Vocabulario Congo: El bantú que se habla en Cuba. Miami: Peninsular Publishing, 1984.

Camp, Stephanie M. H. Closer to Freedom: Enslaved Women and Everyday Resistance in the Plantation South. Chapel Hill: University of North Carolina Press, 2004.

Campbell, Mavis C. The Maroons of Jamaica, 1655–1796: A History of Resistance, Collaboration, and Betrayal. Granby, Mass.: Bergin & Garvey, 1988.

Campion-Vincent, Véronique. "L'image du Dahomey dans la presse française (1890–1895): les sacrifices humaines." Cahiers d'Etudes Africaines 7:25 (1967). Online at: http://etudesafricaines.revues.org.

Carbonell, Walterio. "Plácido, ¿Conspirador?" Revolución y Cultura 2 (1987): 57–58.

Carneiro, Edison. O quilombo dos Palmares, 1630–1695. São Paulo: Brasiliense, 1947.

Caron, Aimery, and Arnold Highfield. The French Intervention in the St. John Slave Revolt of 1733–34. St. Thomas, V.I.: Bureau of Libraries, Museums, and Archaeological Services, Dept. of Conservation and Cultural Affairs, 1981.

Cepero Bonilla, Raúl. Azúcar y abolición. Havana: Ciencias Sociales, 1961.

Chalhoub, Sidney. "Slaves, Freedmen, and the Politics of Freedom in Brazil: The Experience of Blacks in the City of Rio de Janeiro." Slavery & Abolition 10:3 (1989): 64–84.

———. Visões da liberdade: As últimas dêcadas da escravidão na corte. São Paulo: Companhia das Letras, 1990.

Chambers, Douglas B. "My Own Nation: Igbo Exiles in the Diaspora." *Slavery & Abolition* 18:1 (1997): 72–97.

Childs, Matt D. *The 1812 Aponte Rebellion in Cuba and the Struggle against Atlantic Slavery.* Chapel Hill: University of North Carolina Press, 2006.

——. "The Defects of Being a Black Creole: The Degrees of African Identity in the Cuban Cabildos de Nacion, 1790–1820. In *Slaves, Subjects, and Subversives,* ed. Landers and Robinson. 209–46.

Clarke, Hyde. "On the Relations of Culture of the Ashantees." *Journal of the Anthropological Institute of Great Britain and Ireland* 4 (1875): 122–26.

Comaroff, John, and Jean Comaroff. *Ethnography and the Historical Imagination.* Boulder: Westview Press, 1992.

Conrad, Robert. *The Destruction of Brazilian Slavery, 1850–1888.* Berkeley: University of California Press, 1973.

Corbett, Bob. "Introduction to Voodoo in Haiti." Online at: www.webster.edu/~corbetre/haiti/voodoo/overview.htm.

Costa, Emilia Viotti da. *Crowns of Glory, Tears of Blood: The Demerara Slave Rebellion of 1823.* Oxford: Oxford University Press, 1994.

Cotton, J. C. "The People of Old Calabar." *Journal of the Royal African Society* 4:15 (1905): 302–6.

Courlander, Harold. "Gods of the Haitian Mountains." *Journal of Negro History* 29:3 (1944): 339–72.

Craton, Michael. "Proto-Peasant Revolts? The Late Slave Rebellions in the British West Indies, 1816–1832." *Past and Present* 85 (1979): 99–125.

——. "The Passion to Exist: Slave Rebellions in the British West Indies, 1629–1832." *Journal of Caribbean History* 13 (1980): 1–20.

——. *Testing the Chains: Resistance to Slavery in the British West Indies.* Ithaca: Cornell University Press, 1982.

Creswick, H. C. "Life amongst the Veys." *Transactions of the Ethnological Society of London* 6 (1868): 354–61.

Crocker, Lester G. "The Discussion of Suicide in the Eighteenth Century." *Journal of the History of Ideas* 13:1 (1952): 47–72.

Crowder, Michael. *The Story of Nigeria.* London: Faber and Faber, 1962.

Curtin, Philip D. *Economic Change in Precolonial Africa.* Madison: University of Wisconsin Press, 1975.

Daaku, Kwame Yeboah. *Trade and Politics on the Gold Coast, 1600–1720.* Oxford: Clarendon Press, 1970.

Davies, K. G. *The Royal African Company.* London: Longmans, Greens, 1957.

Davis, David Brion. *The Problem of Slavery in Western Culture.* Ithaca: Cornell University Press, 1966.

———. *The Problem of Slavery in the Age of Revolution.* Ithaca: Cornell University Press, 1976.

Davis, Natalie Zemon. "History's Two Bodies." *American Historical Review* 93:1 (1988): 1–30.

Davis, Thomas J. "Conspiracy and Credibility: Look Who's Talking about What—Law Talk and Loose Talk." *William and Mary Quaterly* 3d. ser., 59:1 (2002): 167–74.

Deren, Maya. *The Voodoo Gods.* London: Paladin, 1975.

Deschamps Chapeaux, Pedro. *El negro en la economía habanera del siglo XIX.* Havana: UNEAC, 1971.

———. *Los batallones de pardos y morenos libres.* Havana: Instituto Cubano del Libro, 1976.

———. "Etnias africanas en las sublevaciones de los esclavos en Cuba." *Revista Cubana de Ciencias Sociales* 10:4 (1986). Online at: http://freeweb.supereva.com.

Diaz, Maria Elena. *The Virgin, the King, and the Royal Slaves of El Cobre: Negotiating Freedom in Colonial Cuba, 1670–1780.* Stanford: Stanford University Press, 2002.

Dubois, Laurent. *A Colony of Citizens: Revolution and Slave Emancipation in the French Caribbean, 1787–1804.* Chapel Hill: University of North Carolina Press, 2004.

Durkheim, Émile. *Le suicide. Étude de sociologie.* Paris: F. Alcan, 1897.

Elkins, Stanley M. *Slavery: A Problem in American Institutional and Intellectual Life.* Chicago: University of Chicago Press, 1959.

Elliot, G. F. Scott. "Some Notes on Native West African Customs." *Journal of the Anthropological Institute of Great Britain and Ireland* 23 (1894): 80–83.

Eltis, David, Stephen D. Behrendt, David Richardson, and Herbert Klein, eds. *The Transatlantic Slave Trade: A Database on CD-ROM.* Cambridge: Cambridge University Press, 1999.

Engerman, Stanley, Seymour Drescher, and Robert Paquette, eds. *Slavery.* Oxford: Oxford University Press, 2001.

Entralgo, Elías. "Los problemas de la esclavitud: la conspiración de Aponte." *Cuadernos de Historia Habanera* 12 (1937).

Estévez y Romero, Luis. *Desde el Zanjón hasta Baire: Datos para la historia política de Cuba.* Havana: Imprenta La Propaganda Literaria, 1899.

Eyre, L. A. "The Maroon Wars in Jamaica: A Geographical Appraisal." *Jamaican Historical Review* 12 (1980): 80–102.

Falola, Toyin. *Yoruba Gurus: Indigenous Production of Knowledge in Africa.* Trenton, N.J.: Africa World Press, 1999.

———, and Michael R. Doortmont. "Iwe Itan Oyo: A Traditional Yoruba History and Its Author." *Journal of African History* 30:2 (1989): 301–29.

———, and Matt D. Childs, eds. *The Yoruba Diaspora in the Atlantic World.* Bloomington: Indiana University Press, 2004.

Feijóo, Samuel. *El negro en la literature folklorica cubana.* Havana: Letras Cubanas, 1980.

Fentress, James, and Chris Wickham. *Social Memory.* Oxford: Blackwell, 1992.

Ferreira, Jackson. "Por hojese acaba a lida: suicidio escravo na Bahia (1850–1888)." Afro-Asia 31 (2004): 197–234.

Ferrer, Ada. "Noticias de Haití en Cuba." Revista de Indias 63:229 (2003): 675–94.

Fick, Carolyn E. The Making of Haiti: The Saint Domingue Revolution from Below. Knoxville: University of Tennessee Press, 1989.

———. "The Saint Domingue Slave Insurrection of 1791: A Socio-Political and Cultural Analysis." Journal of Caribbean History 25:1 & 2 (1991): 1–40.

Fields, Karen E. "Charismatic Religion as Popular Protest: The Ordinary and the Extraordinary in Social Movements." Theory and Society 11:3 (1982): 321–61.

Finkleman, Paul, ed. Slavery and the Law. Madison: University of Wisconsin Press, 1997.

Flanigan, Daniel J. "Criminal Procedure in Slave Trials in the Antebellum South." Journal of Southern History 40:4 (1974): 537–64.

Fox, Richard G., and Orin Starn, eds. Between Resistance and Revolution: Cultural Politics and Social Protest. New Brunswick, N.J.: Rutgers University Press, 1997.

Franco, José Luciano. La conspiración de Aponte. Havana: Publicaciones del Archivo Nacional de Cuba, 1963.

———. Esclavitud, comercio y tráfico negreros. Havana: Archivo Nacional de Cuba, 1974.

———. Contrabando y trata negrera en el Caribe. Havana: Ciencias Sociales, 1976.

———. Comercio clandestino de esclavos. Havana: Ciencias Sociales, 1980.

Freitas, Décio. Palmares, a guerra dos escravos. Rio de Janeiro: Graal, 1982.

Freyre, Gilberto. The Masters and the Slaves [Casa-Grande & Senzala]: A Study in the Development of Brazilian Civilization. Berkeley: University of California Press, 1986.

Fuente, Alejandro de la. A Nation for All: Race, Inequality, and Politics in Twentieth-Century Cuba. Chapel Hill: University of North Carolina Press, 2001.

———. "Slave Law and Claims-Making in Cuba: The Tannenbaum Debate Revisited." Law and History Review 22:2 (2004): 339–69.

———. "Su 'único derecho': Los esclavos y la ley." Debates y Perspectivas: Cuadernos de historia y ciencias sociales 4 (2004): 7–22.

———. "La esclavitud, la ley, y la reclamación de derechos en Cuba: Repensando el debate de Tannenbaum." Debates y Perspectivas: Cuadernos de historia y ciencias sociales 4 (2004): 37–68.

Funes Monzote, Reinaldo. De Bosque a sabana: Azúcar, deforestación, y medio ambiente en Cuba, 1492–1926. Mexico, D.F.: Siglo XXI, 2004.

García, Enildo A. Cuba: Plácido, poeta mulato de la emancipación, 1809–1844. New York: Senda Nueva Ediciones, 1986.

García, Gloria. "A propósito de La Escalera: El esclavo como sujeto político." Boletín del Archivo Nacional de Cuba 12 (2000): 1–13.

———. La esclavitud desde la esclavitud: La visión de los siervos. Havana: Ciencias Sociales, 2003.

Gaspar, David Barry. Bondmen and Rebels: A Study of Master-Slave Relations in Antigua. Baltimore: Johns Hopkins University Press, 1985.

———, and David P. Geggus, eds. *A Turbulent Time: The French Revolution and the Greater Caribbean*. Bloomington: Indiana University Press, 1997.

Geggus, David P. "The Bois Caïman Ceremony." *Journal of Caribbean History* 25:1 & 2 (1991): 41–57.

———. "Haitian Voodoo in the Eighteenth Century: Language, Culture, and Resistance." *Jahrbuch für Geschichte von Staat, Wirtschaft, und Gesellschaft Latein Amerikas* 28 (1991): 21–51.

———. "Slave Resistance in the Spanish Caribbean in the Mid-1790s." In *A Turbulent Time*, ed. Gaspar and Geggus. 131–55.

———. "Slavery, War, and Revolution in the Greater Caribbean, 1789–1815." In *A Turbulent Time*, ed. Gaspar and Geggus. 1–50.

———, ed. *The Impact of the Haitian Revolution in the Atlantic World*. Columbia: University of South Carolina Press, 2001.

Gemery, Henry. A., and Jan. S. Hogendorn, eds. *The Uncommon Market: Essays in the Economic History of the Atlantic Slave Trade*. New York: Academic Press, 1979.

Genovese, Eugene. *Roll, Jordan, Roll: The World the Slaves Made*. New York: Pantheon, 1974.

———. *From Rebellion to Revolution: Afro-American Slave Revolts in the Making of the Modern World*. Baton Rouge: Louisiana State University Press, 1979.

Ginzburg, Carlo. *The Cheese and the Worms: The Cosmos of a Sixteenth-Century Miller*. Baltimore: Johns Hopkins University Press, 1980.

———. "Microhistory: Two or Three Things That I Know about It." *Critical Inquiry* 20 (1993): 10–35.

———, and Carlo Poni. "Il nome e il come: scambio ineguale e mercato storiografico." *Quaderni Storici* 40 (1979): 181–90.

Gledhill, John. *Power and Its Disguises: Anthropological Perspectives on Politics*. London: Pluto Press, 1994.

Gluckman, Max. *Custom and Conflict in Africa*. Oxford: Blackwell, 1955.

Goffman, Erving. *The Presentation of Self in Everyday Life*. Garden City, N.Y.: Doubleday, 1958.

———. *Encounters: Two Studies in the Sociology of Interaction*. Indianapolis: Bobbs-Merrill, 1961.

———. "The Interaction Order: American Sociological Association, 1982 Presidential Address." *American Sociological Review* 48:1 (1983): 1–17.

Gomes, Flavio dos Santos. "Quilombos do Rio de Janeiro no século XIX." In *Liberdade por um fio*, ed. Reis and Gomes. 263–90.

Gomez, Michael. *Exchanging Our Country Marks: The Transformation of African Identities in the Colonial and Antebellum South*. Chapel Hill: University of North Carolina Press, 1998.

————. "African Slavery and Identity in the Americas." *Radical History Review* 75 (1999): 111–20.

González del Valle, Francisco. *La conspiración de La Escalera*. Havana: El Siglo XX, 1925.

Gottlieb, Roger. "The Concept of Resistance: Jewish Resistance during the Holocaust." *Social Theory and Practice* 9:1 (1983): 31–49.

Goveia, Elsa V. *Slave Society in the British Leeward Islands at the End of the Eighteenth Century*. New Haven: Yale University Press, 1965.

————. *The West Indian Slave Laws of the Eighteenth Century*. Kingston: Caribbean University Press, 1970.

Gregory, Steven. *Santería in New York City: A Study in Cultural Resistance*. New York: Routledge, 1999.

Grinberg, Keila. *Liberata. A lei da ambiguidade. As ações de liberdade da Corte de Apelação do Rio de Janeiro no século XIX*. Rio de Janeiro: Relume Dumara, 1994.

————. "La manumission, el género, y la ley en el Brasil del siglo XIX: El proceso legal de Liberata por su libertad." *Debates y Perspectivas: Cuadernos de historia y ciencias sociales* 4 (2004): 89–104.

Guerra, Ramiro. *Manual de historia de Cuba*. Havana: Ciencias Sociales, 1971.

Gutmann, Matthew. "Rituals of Resistance: A Critique of the Theory of Everyday Forms of Resistance." *Latin American Perspectives* 20:2 (1993): 74–92.

Haley, Alex. *Roots*. London: Hutchinson, 1977.

Hall, Gwendolyn Midlo. *Social Control in Slave Plantation Societies: A Comparison of St. Domingue and Cuba*. Baltimore: Johns Hopkins University Press, 1971.

————. *Slavery and African Ethnicities in the Americas: Restoring the Links*. Chapel Hill: University of North Carolina Press, 2005.

Harris, J. M. "Some Remarks on the Origin, Manners, Customs, and Superstitions of the Callinas People of Sierra Leone." *Journal of the Anthropological Society of London* 4 (1866): lxxxii–lxxxv.

Harris, Joseph E., ed. *Global Dimensions of the African Diaspora*. Washington, D.C.: Howard University Press, 1993.

Hast, Adele. "The Legal Status of the Negro in Virginia, 1705–1765." *Journal of Negro History* 54:3 (1969): 217–39.

Hearn, Adrian. "Transformation, Transcendence, or Transculturation? The Many Faces of Cuban Santería." *Humanities Research Centre* 10: 1 (2003): 56–62.

Helg, Aline. *Our Rightful Share: The Afro-Cuban Struggle for Equality, 1886–1912*. Chapel Hill: University of North Carolina Press, 1995.

Herskovits, Melville. *The Myth of the Negro Past*. New York: Harper Brothers, 1941.

Heywood, Linda M. "The Angola-Afro-Brazilian Cultural Connection." *Slavery & Abolition* 20:1 (1999): 9–23.

Hilton, Anne. *The Kingdom of Kongo*. Oxford: Clarendon Press, 1985.

Hobsbawn, Eric J. *The Age of Revolution, 1789–1848*. Cleveland: World, 1962.

Hoogbergen, Wim. *The Boni Maroon Wars in Suriname*. Leiden: E. J. Brill, 1990.

Howard, Philip. *Changing History: Afro-Cuban Cabildos and Societies of Color in the Nineteenth Century*. Baton Rouge: Louisiana State University Press, 1998.

Huggins, Nathan, Martin Kilson, and Daniel Fox, eds. *Key Issues in the Afro-American Experience*. New York: Harcourt Brace Jovanovich, 1971.

Ibarra Cuesta, Jorge. *Ideología mambisa*. Havana: Instituto del Libro, 1967.

Iduarte, Juan. "Noticias sobre sublevaciones y conspiraciones de esclavos. Cafetal Salvador, 1833." *Revista de la Biblioteca Nacional José Martí* 73:24, 1–2, 3ra época (1982): 117–52.

Inikori, Joseph E. "The Import of Firearms into West Africa, 1750–1807." *Journal of African History* 18:3 (1977): 339–68.

Isaacman, Allen. "Peasants and Rural Social Protest in Africa." *African Studies Review* 33:2 (1990): 1–120.

James, C. L. R. *The Black Jacobins: Toussaint L'Ouverture and the San Domingo Revolution*. New York: Vintage, 1963.

Johnson, Sherry. *The Social Transformation of Eighteenth-Century Cuba*. Gainesville: University Press of Florida, 2001.

Johnson, Walter, ed. *The Chattel Principle: Internal Slave Trades in the Americas, 1808–1888*. New Haven: Yale University Press, 2004.

Jones, Adam. "Who Were the Vai?" *Journal of African History* 22:2 (1981): 159–78.

Jordan, Winthrop D. *Tumult and Silence at Second Creek: An Inquiry into a Civil War Slave Conspiracy*. Baton Rouge: Louisiana State University Press, 1993.

Joseph, Gilbert M. "On the Trail of Latin American Bandits." *Latin American Research Review* 25:3 (1990): 7–18.

Kafka, Franz. *The Trial*. Trans. Willa and Edwin Muir. New York: Schocken Books, 1995.

Karasch, Mary. *Slave Life in Rio de Janeiro, 1808–1850*. Princeton: Princeton University Press, 1987.

Kea, R. A. "Firearms and Warfare on the Gold and Slave Coasts from the Sixteenth to the Nineteenth Centuries." *Journal of African History* 12:2 (1971): 185–213.

Kelly, Robin D. G. *Race Rebels: Culture, Politics, and the Black Working Class*. New York: Free Press, 1994.

Kent, Raymond. "Palmares, An African State in Brazil." *Journal of African History* 6:2 (1965): 161–75.

Kiple, Kenneth. *The Caribbean Slave: A Biological History*. Cambridge: Cambridge University Press, 1984.

———. "Cholera and Race in the Caribbean." *Journal of Latin American Studies* 17:1 (1985): 157–77.

———, and Virginia Himmelsteib King. *Another Dimension to the African Diaspora: Diet, Disease, and Racism*. Cambridge: Cambridge University Press, 1981.

Klein, Herbert S. Slavery in the Americas: A Comparative Study of Virginia and Cuba. Chicago: University of Chicago Press, 1967.

Knight, Franklin. Slave Society in Cuba during the Nineteenth Century. Madison: University of Wisconsin Press, 1970.

Kokotovic, Misha. The Colonial Divide in Peruvian Narrative: Social Conflict and Transculturation. Brighton: Sussex Academic Press, 2005.

Kopytoff, Barbara K. "Colonial Treaty as Sacred Charter of the Jamaican Maroons." Ethnohistory 26:1 (1979): 45–64.

Koslow, Philip. Yorubaland: The Flowering of Genius. New York: Chelsea House, 1996.

Kuethe, Allan J. Cuba, 1753–1815: Crown, Military, and Society. Knoxville: University of Tennessee Press, 1986.

Kundera, Milan. The Joke. Trans. Michael Henry Heim. Harmondsworth: Penguin, 1983.

———. The Unbearable Lightness of Being. Trans. Michael Henry Heim. New York: Harper-Perennial, 1991.

Kup, A. P. A History of Sierra Leone. Cambridge: Cambridge University Press, 1962.

Kutzinski, Sugar's Secrets: Race and the Erotics of Cuban Nationalism. Charlottesville: University Press of Virginia, 1993.

Landers, Jane, and Barry Robinson, eds. Slaves, Subjects, and Subversives: Blacks in Colonial Latin America. Albuquerque: University of New Mexico Press, 2006.

La Rosa Corzo, Gabino. Los cimarrones de Cuba. Havana: Ciencias Sociales, 1988.

———. Runaway Slave Settlements in Cuba: Resistance and Repression. Trans. Mary Todd. Chapel Hill: University of North Carolina Press, 2003.

———, and Mirtha T. González, eds. Cazadores de esclavos: Diarios. Havana: Fundación Fernando Ortiz, 2004.

Latham, A. J. H. "Witchcraft Accusations and Economic Tensions in Pre-Colonial Old Calabar." Journal of African History 13:2 (1972): 249–60.

Law, Robin. "The Constitutional Troubles of Oyo in the Eighteenth Century." Journal of African History 12:1 (1971): 25–44.

———. "A West African Cavalry State: The Kingdom of Oyo." Journal of African History 16:1 (1975): 1–15.

———. The Oyo Empire c. 1600–c. 1836: A West African Imperialism in the Era of the Atlantic Slave Trade. Oxford: Oxford University Press, 1977.

———. The Horse in West African History. Oxford: Oxford University Press, 1980.

———. "Human Sacrifice in Pre-Colonial West Africa." African Affairs 84:334 (1985): 53–87.

———. The Slave Coast of West Africa, 1550–1750: The Impact of the Atlantic Slave Trade on an African Society. Oxford: Clarendon Press, 1991.

———. "Ethnicity and the Slave Trade: 'Lucumi' and 'Nago' as Ethnonyms in West Africa." History in Africa 24 (1997): 205–19.

————. "On the African Background to the Slave Insurrection in Saint-Domingue (Haïti) in 1791: The Bois Caiman Ceremony and the Dahomian 'Blood Pact.'" Paper presented at the Harriet Tubman Seminar. Department of History. York University. Monday, 8 November 1999.

Lears, T. J. Jackson. "The Concept of Cultural Hegemony: Problems and Possibilities." *American Historical Review* 90:3 (1985): 567–93.

Legrand, Catherine C. "Informal Resistance on a Dominican Sugar Plantation during the Trujillo Dictatorship." *Hispanic American Historical Review* 75:4 (1995): 555–96.

Lewis, Maureen Warner. *Guinea's Other Suns: The African Dynamic in Trinidad Culture.* Dover, Mass.: Majority Press, 1991.

————. *Trinidad Yoruba: From Mother Tongue to Memory.* Tuscaloosa: University of Alabama Press, 1996.

Link, William A. *Roots of Secession: Slavery and Politics in Antebellum Virginia.* Chapel Hill: University of North Carolina Press, 2003.

López Denis, Adrián. "El cólera en la Habana: La epidemia de 1833." M.A. thesis, University of Havana, 2000.

————. "Higiene pública contra higiene privada: Cólera, limpieza, y poder en la Habana colonial." *Estudios Interdisciplinarios de América Latina y el Caribe* 14:1 (2003): 11–33.

Lovejoy, Paul. *Transformations in Slavery: A History of Slavery in Africa.* Cambridge: Cambridge University Press, 1983.

————. "The African Diaspora: Revisionist Interpretations of Ethnicity, Culture, and Religion under Slavery." *Studies in the World History of Slavery, Abolition, and Emancipation* 2:1 (1997). www2.h-net.msu.edu/~slavery/essays/esy9701love.html.

————. *Slavery, Commerce, and Production in the Sokoto Caliphate of West Africa.* Trenton, N.J.: Africa World Press, 2005.

Lucena Salmoral, Manuel. *Los códigos negros de la América Española, 1768–1842.* Alcalá de Henares: UNESCO/Universidad de Alcalá de Henares, 2003.

Magee, Brian. *The English Recusants.* London: Burns, Oates, and Washbourne, 1938.

Mallon, Florencia. *Peasant and Nation: The Making of Postcolonial México and Perú.* Berkeley: University of California Press, 1995.

Mann, Kristin. "Shifting Paradigms in the Study of the African Diaspora and of Atlantic History and Culture." *Slavery & Abolition* 22:1 (2001): 3–22.

Marquese, Rafael de Bivar. *Idéias sobre a administração das plantations escravistas nas Américas, séc. XVII–XIX.* Relatório de Pesquisa no. 4. São Paulo: FAPESP, 2000.

Márquez, José de Jesús. "Conspiración de Aponte." *Revista Cubana* 19 (1894): 441–54.

————. "Plácido y los conspiradores de 1844." *Revista Cubana* 20 (1894): 35–51.

Marrero, Leví. *Cuba: Economía y sociedad.* 13 vols. Madrid: Playor, [1981].

Martínez Alier, Verena. *Marriage, Class, and Colour in Nineteenth-Century Cuba: A Study of Racial Attitudes and Sexual Values in a Slave Society.* Cambridge: Cambridge University Press, 1974.

Martínez García, Daniel. "La sublevación de la Alcancía: Su rehabilitación histórica en el proceso conspirativo que concluye en La Escalera (1844)." Rábida 19 (2000): 41–48.

Mattos, Hebe. Das cores do silêncio: Os significados da liberdade no sudeste escravista—Brasil, século XIX. Rio de Janeiro: Nova Fronteira, 1998.

Meyer, Jean. L'armament nantais dans la deuxième moitié du XVIIIe siècle. Paris: SEVPEN, 1969.

Miller, Ivor L. "Religious Symbolism in Cuban Political Performance." Drama Review 44: 2 (2000): 30–55.

Miller, Joseph C. Way of Death: Merchant Capitalism and the Angolan Slave Trade. Madison: University of Wisconsin Press, 1988.

Mintz, Sidney W., and Richard Price. An Anthropological Approach to the Afro-American Past: A Caribbean Perspective. Philadelphia: ISHI, 1976.

Montalvo J. R., C. de la Torre, and L. Montané. El cráneo de Maceo. Estudio antropológico. Havana: Imprenta Militar, 1900.

Montilus, Guérin. Dieux en diaspora: Les loa haïtiens et les vaudou du royaume d'Allada (Bénin). Niamey: CELHTO, 1988.

———."Guinea versus Congo Lands: Aspects of the Collective Memory in Haiti." In Global Dimensions of the African Diaspora, ed. Harris. 159–65.

Moore, Barrington, Jr. Injustice: The Social Bases of Obedience and Revolt. White Plains, N.Y.: M. E. Sharpe, 1978.

Moore, Wilbert E. "Slave Law and the Social Structure." Journal of Negro History 26:2 (1941): 171–202.

Morales y Morales, Vidal. Iniciadores y primeros mártires de la revolución cubana. Havana: Consejo Nacional de Cultura, 1963.

Moreno Fraginals, Manuel. "Africa in Cuba: A Quantitative Analysis of the African Populations in the Island of Cuba." In Comparative Perspectives on Slavery in New World Plantation Societies, ed. Rubin and Tuden. 187–201.

———. El ingenio: Complejo económico social cubano del azúcar. 3 vols. Havana: Ciencias Sociales, 1978.

———, Herbert S. Klein, and Stanley L. Engerman. "The Level and Structure of Slave Prices on Cuban Plantations in the Mid-Nineteenth Century: Some Comparative Perspectives." American Historical Review 88:5 (1983): 1201–18.

Morgan, Philip D. "The Cultural Implications of the Atlantic Slave Trade: African Regional Origins, American Destinations, and New World Developments." Slavery & Abolition 18:1 (1997): 122–45.

Morris, Christopher. "The Articulation of Two Worlds: The Master-Slave Relationship Reconsidered." Journal of American History 85:3 (1998): 982–1007.

Mortimore, M. J. "Population Densities and Systems of Agricultural Land Use in Northern Nigeria." Nigerian Geographical Journal 14 (1971): 3–15.

Muir, Edward, and Guido Ruggiero, eds. *Microhistory and the Lost Peoples of Europe: Selections from Quaderni Storici.* Baltimore: Johns Hopkins University Press, 1991.

Mullin, Michael. *Africa in America: Slave Acculturation in the American South and the British Caribbean, 1736–1831.* Urbana: University of Illinois Press, 1992.

Murray, David. *Odious Commerce: Britain, Spain, and the Abolition of the Cuban Slave Trade.* Cambridge: Cambridge University Press, 1980.

———. "The Slave Trade, Slavery, and Cuban Independence." *Slavery & Abolition* 20:3 (1999): 106–26.

Nelson, Cary, and Lawrence Grossberg, eds. *Marxism and the Interpretation of Culture.* Urbana: University of Illinois Press, 1988.

Nodal, Roberto, Rolando A. Alum, and Rafael Núñez. *Linguistic Folklore in the Latin Caribbean: A Selected Glossary of the Abakuá Language in Cuba.* Milwaukee: University of Wisconsin–Milwaukee, 1977.

Noon, Gloria. "On Suicide." *Journal of the History of Ideas* 39:3 (1978): 371–86.

Nordstrom, Carl. "The New York Slave Code." *Afro-Americans in New York Life and History* 4:1 (1980): 7–26.

Núñez-Cedeño, Rafael A., Roberto Nodal, and Rolando A. Alúm. "The Afro-Hispanic Abakuá: A Study of Linguistic Pidginization." *Orbis* 31: 1 & 2 (1985): 263–84.

Nuruddin, Yusuf. "The Sambo Thesis Revisited: Slavery's Impact upon the African American Personality." *Socialism and Democracy* 17:2 (2003). Online at: www.sdonline.org.

Offiong, Daniel A. "The Status of Slaves in Igbo and Ibibio in Nigeria." *Phylon* 46:1 (1985): 49–57.

O'Hanlon, Rosalind. "Recovering the Subject: Subaltern Studies and Histories of Resistance in Colonial South Asia." *Modern Asian Studies* 22:1 (1988): 189–224.

O'Hear, Ann. *Power Relations in Nigeria: Ilorin Slaves and Their Successors.* Rochester, N.Y.: University of Rochester Press, 1997.

Ortiz, Fernando. *Los negros brujos.* Madried: Libreria de Fernando Fe, 1906.

———. *Los negros esclavos.* Havana: Bimestre Cubana, 1916.

———. *Contrapunteo cubano del tabaco y el azúcar.* Caracas: Biblioteca Ayacucho, 1947.

———. *Los instrumentos de la música afrocubana.* Havana: Cárdenas y Cía, 1954.

———. *Historia de una pelea cubana contra los demonios.* Havana: Ucar, Garcia, 1959.

———. *Los bailes y el teatro de los negros en Cuba.* Havana: Letras Cubanas, 1981.

———. *Ensayos Etnográficos.* Havana: Ciencias Sociales, 1984.

———. *Glosario de Afronegrismos.* Havana: Ciencias Sociales, 1990.

Ortner, Sherry B. "Resistance and the Problem of Ethnographic Refusal." *Comparative Studies in Society and History* 37:1 (1995): 173–93.

Orwell, George. *1984.* London: Secker & Warburg, 1949.

Palmer, R. R. *The Age of Democratic Revolution: A Political History of Europe and America, 1760–1800.* 2 vols. Princeton: Princeton University Press, 1959.

Palmié, Stephan. *Wizards and Scientists: Explorations in Afro-Cuban Modernity and Tradi-tion.* Durham, N.C.: Duke University Press, 2002.

Paquette, Robert L. *Sugar Is Made with Blood: The Conspiracy of La Escalera and the Con-flict between Empires over Slavery in Cuba.* Middletown, Conn.: Wesleyan University Press, 1987.

———. "Social History Update: Slave Resistance and Social History." *Journal of Social History* 24:3 (1991): 681–85.

———. "The Drivers Shall Lead Them: Image and Reality in Slave Resistance." In *Slav-ery, Secession, and Southern History,* ed. Paquette and Ferleger. 31–58.

———. "Jacobins of the Lowcountry: The Vesey Plot on Trial." *William and Mary Quar-terly* 3d. ser., 59:1 (2002): 185–92.

———, and Joseph C. Dorsey. "The Escoto Papers and Cuban Slave Resistance." *Slavery & Abolition* 15:3 (1994): 88–95.

——— and Douglas R. Egerton. "On Facts and Fables: New Light on the Denmark Vesey Affair." *South Carolina Historical Magazine* 105:1 (2004): 8–48.

———, and Louis A. Ferleger, eds. *Slavery, Secession, and Southern History.* Charlottesville: University Press of Virginia, 2000.

Patterson, Orlando. *The Sociology of Slavery: An Analysis of the Origins, Development, and Structure of Negro Slave Society in Jamaica.* Rutherford, N.J.: Fairleigh Dickinson Press, 1967.

———. *Slavery and Social Death: A Comparative Study.* Cambridge, Mass.: Harvard Uni-versity Press, 1982.

Paulson, Timothy J. *Days of Sorrow, Years of Glory, 1813–1850: From the Nat Turner Revolt to the Fugitive Slave Law.* New York: Chelsea House, 1994.

Pearson, Edward A. "Trials and Errors: Denmark Vesey and His Historians." *William and Mary Quarterly* 3d. ser., 59:1 (2002): 137–42.

Peel, J. D. Y. *Religious Encounters and the Making of the Yoruba.* Bloomington: Indiana University Press, 2000.

Penalves Rocha, Antonio. *A economia política na sociedade escravista.* São Paulo: HU-CITEC, 1996.

Perez, Louis A. *To Die in Cuba: Suicide and Society.* Chapel Hill: University of North Carolina Press, 2005.

Pérez de la Riva, Juan. *El barracón y otros ensayos.* Havana: Ciencias Sociales, 1975.

Phillips, Ulrich Bonnell. *Life and Labor in the Old South.* Boston: Little, Brown, 1929.

Piersen, William D. "White Cannibals, Black Martyrs: Fear, Depression, and Religious Faith as Causes of Suicide among New Slaves." *Journal of Negro History* 62:2 (1977): 147–59.

Pope-Hennessy, James. *Sins of the Fathers: A Study of the Atlantic Slave Traders, 1441–1807.* London: Weidenfeld and Nicholson, 1967.

Poumier Taquechel, María. "El suicidio esclavo en Cuba en los años 1840." Anuario de Estudios Americanos 43 (1986): 69–86.

Pratt, M. L. Imperial Eyes: Travel Writing and Transculturation. London: Routledge, 1992.

Price, Richard. Maroon Societies: Rebel Slave Communities in the Americas. Garden City, N.Y.: Anchor Press, 1973.

———. The Guiana Maroons: A Historical and Bibliographical Introduction. Baltimore: Johns Hopkins University Press, 1976.

Prince, Howard M. "Slave Rebellion in Bahia, 1807–1835." Ph.D. diss., Columbia University, 1972.

Quiñones, Tato. Ecorie Abakuá: Cuatro ensayos sobre los ñáñigos cubanos. Havana: Ediciones Unión, 1994.

Rasnake, Roger N. Domination and Cultural Resistance: Authority and Power among an Andean People. Durham, N.C.: Duke University Press, 1988.

Reis, João José. Slave Rebellion in Brazil: The Muslim Uprising of 1835 in Bahia. Baltimore: Johns Hopkins University Press, 1993.

———. "Escravos e coiteiros no quilombo do Oitizeiro." In Liberdade por um fio, ed. Reis and Gomes. 332–72.

———. "The Revolution of the Ganhadores: Urban Labor, Ethnicity, and the African Strike of 1857 in Bahia, Brazil." Journal of Latin American Studies 29:2 (1997): 355–93.

———. Death Is a Festival: Funeral Rites and Rebellion in Nineteenth-Century Brazil. Chapel Hill: University of North Carolina Press, 2003.

———. "Batuque: African Drumming and Dance between Repression and Concession, Bahia, 1808–1855." Bulletin of Latin American Studies 24:2 (2005): 201–14.

———, and Eduardo Silva. Negociação e conflito: A resistência negra no Brasil escravista. São Paulo: Companhia das Letras, 1999.

———, and Flavio dos Santos Gomes, eds. Liberdade por um fio: História dos quilombos no Brasil. São Paulo: Companhia das Letras, 1996.

Richards, P. Indigenous Agricultural Revolution: Ecology and Food Production in West Africa. London: Hutchinson, 1985.

Richards, W. A. "The Import of Firearms into West Africa in the Eighteenth Century." Journal of African History 21:1 (1980): 43–59.

Richardson, David. "West African Consumption Patterns and Their Influence on the Eighteenth-Century English Slave Trade." In The Uncommon Market: Essays in the Economic History of the Atlantic Slave Trade, ed. Gemery and Hogendorn. 303–30.

Rodney, Walter. A History of the Upper Guinea Coast. Oxford: Clarendon Press, 1970.

Röhrig Assunção, Matthias. "Quilombos maranhaenses." In Liberdade por um fio, ed. Reis and Gomes. 433–66.

———. Capoeira: The History of an Afro-Brazilian Martial Art. London: Routledge, 2005.

———. "Brazilian Popular Culture, or the Curse and Blessings of Cultural Hybridism." Bulletin of Latin American Studies 24:2 (2005): 157–67.

Rosenhaft, Eve. "History, Anthropology, and the Study of Everyday Life." *Comparative Studies in Society and History* 29 (1987): 99–105.

Roura Alvarez, Lissette. "Tercera campaña arqueológica en el antiguo cafetal El Padre." *Boletín del Gabinete de Arqueología* 2:2 (2002): 133–34.

Rubin, Vera, and Arthur Tudin, eds. *Comparative Perspectives on Slavery in New World Plantation Societies.* New York: New York Academy of Sciences, 1977.

Rudé, George. *The Crowd in History: A Study of Popular Disturbances in France and England, 1730–1848.* London: Lawrence and Wishart, 1981.

Sala-Molins, Louis. *L'Afrique aux Amériques. Le Code Noir espagnol.* Paris: Presses Universitaires de France, 1992.

Sanabria, Harry. "Resistance and the Arts of Domination: Miners and the Bolivian State." *Latin American Perspectives* 27:1 (2000): 56–81.

Sánchez, Juan. "'José Dolores, capitán de cimarrones.' Un capítulo inédito de las rebeldías de esclavos en Matanzas." *Bohemia* 66 (November 1974): 50–53.

Sanguily, Manuel. *José de la Luz y Caballero (estudio crítico).* 1890. Reprint, Havana: Consejo Nacional de Cultura, 1962.

———. "Un improvisador cubano (el poeta Plácido y el juicio de Menéndez Pelayo)." *Hojas Literarias* 3 (1894): 93–120.

———. "Otra vez Plácido y Menéndez Pelayo." *Hojas Literarias* 3 (1894): 227–254.

———. "Una opinión en contra de Plácido (notas críticas)." *Hojas Literarias* 4 (1894): 425–35.

Sarracino, Rodolfo. *Inglaterra: Sus dos caras en la lucha cubana por la abolición.* Havana: Letras Cubanas, 1989.

Schafer, Judith Kelleher. *Becoming Free, Remaining Free: Manumission and Enslavement in New Orleans, 1846–1862.* Baton Rouge: Louisiana State University Press, 2003.

Schuler, Monica. "Akan Slave Rebellions in the British Caribbean." *Savacou* 1:1 (1970): 8–31.

———. "Day-to-Day Resistance to Slavery in the Caribbean during the Eighteenth Century." *Bulletin of African Studies Association of the West Indies* 6 (1973): 57–75.

Schwartz, Stuart. "The Manumission of Slaves in Colonial Brazil: Bahia, 1684–1745." *Hispanic American Historical Review* 54 (1974): 603–35.

———. "Resistance and Accommodation in Eighteenth-Century Brazil: The Slaves' View of Slavery." *Hispanic American Historical Review* 57 (1977): 69–81.

———. *Sugar Plantations in the Formation of Brazilian Society: Bahia, 1550–1835.* Cambridge: Cambridge University Press, 1985.

Scott, James C. *Weapons of the Weak: Everyday Forms of Peasant Resistance.* New Haven: Yale University Press, 1985.

———. *Domination and the Arts of Resistance: Hidden Transcripts.* New Haven: Yale University Press, 1990.

Scott, Julius. "The Common Wind: Currents of Afro-American Communication in the Era of the Haitian Revolution." Ph.D. diss., Duke University, 1986.

Scott, Rebecca. *Slave Emancipation in Cuba: The Transition to Free Labor, 1860–1899.* Princeton: Princeton University Press, 1985.

Shepherd, Verene, and Glen Richards, eds. *Questioning Creole: Creolisation Discourses in Caribbean Culture.* Kingston: Ian Randle, 2002.

Sheridan, Richard B. "The Guinea Surgeons on the Middle Passage: The Provision of Medical Services in the British Slave Trade." *International Journal of African Historical Studies* 14:4 (1981): 601–25.

Singleton, Theresa. "Slavery and Spatial Dialectics on Cuban Coffee Plantations." *World Archaeology* 33:1 (2001): 98–114.

Sio, Arnold A. "Interpretations of Slavery: The Slave Status in the Americas." *Comparative Studies in Society and History* 7:3 (1965): 289–308.

Slenes, Robert. *Na Senzala, uma flor: Esperanças e recordações na formação da família escrava, Brasil sudeste, século XIX.* Rio de Janeiro: Nova Fronteira, 1999.

Smith, Robert. "Yoruba Armament." *Journal of African History* 8:1 (1967): 87–106.

———. "The Canoe in West African History." *Journal of African History* 11:4 (1970): 515–33.

Sontag, Susan. "The Power of Principle." *The Guardian: Review* (26 April 2003): 4–6.

Soumonni, Elisée. *Daomé e o mundo atlântico.* Amsterdam: SEPHIS, 2001.

Spitta, Silvia. *Between Two Waters: Narratives of Transculturation in Latin America.* College Station: Texas A&M University Press, 2006.

Spivak, Gayatri C. "Can the Subaltern Speak?" In *Marxism and Interpretation of Culture,* ed. Nelson and Grossberg. 271–313.

Stampp, Kenneth M. *The Peculiar Institution: Slavery in the Antebellum South.* New York: Vintage, 1956.

Steckel, Richard H., and Richard A. Jensen. "New Evidence on the Causes of Slave and Crew Mortality in the Atlantic Slave Trade." *Journal of Economic History* 46:1 (1986): 57–77.

Stein, Robert L. *The French Slave Trade in the Eighteenth Century: An Old Regime Business.* Madison: University of Wisconsin Press, 1979.

Stinchcombe, Arthur L. "Freedom and Oppression of Slaves in the Eighteenth-Century Caribbean." *American Sociological Review* 59:6 (1994): 911–27.

Suttles, William C., Jr. "African Religious Survivals as Factors in American Slave Revolts." *Journal of Negro History* 56:2 (1971): 97–104.

Sweet, James H. *Recreating Africa: Culture, Kinship, and Religion in the African-Portuguese World, 1441–1770.* Chapel Hill: University of North Carolina Press, 2003.

———. "Manumission in Rio de Janeiro, 1749–54: An African Perspective." *Slavery & Abolition* 24:1 (2003): 54–70.

Tannenbaum, Frank. *Slave and Citizen: The Negro in the Americas.* New York: Knopf, 1946.

Thornton, John K. *The Kingdom of Kongo: Civil War and Transition.* Madison: University of Wisconsin Press, 1983.

———. "The Art of War in Angola." *Comparative Studies in Society and History* 30:2 (1988): 360–78.

———. "On the Trail of Voodoo: African Christianity in Africa and the Americas." *The Americas* 44:1 (1988): 261–78.

———. "African Soldiers in the Haitian Revolution." *Journal of Caribbean History* 25:1 & 2 (1991): 50–80.

———. "African Dimensions of the Stono Rebellion." *American Historical Review* 96:4 (1991): 1101–13.

———. *Africa and the Africans in the Making of the Atlantic World, 1400–1800.* Cambridge: Cambridge University Press, 1992.

———. "'I Am the Subject of the King of Kongo': African Political Ideology and the Haitian Revolution." *Journal of World History* 4:2 (1993): 181–214.

———. *Warfare in Atlantic Africa, 1500–1800.* London: Routledge, 1999.

Tomish, Dale W. "The Wealth of Empire: Francisco de Arango y Parreño, Political Economy, and the Second Slavery in Cuba." *Comparative Studies in Society and History* 45:1 (2003): 4–28.

———. *Through the Prism of Slavery: Labor, Capital, and World Economy.* Lanham, Md.: Rowman and Littlefield, 2004.

Tushnet, Mark V. *Slave Law in the American South: State v. Mann in History and Literature.* Lawrence: University Press of Kansas, 2003.

Volpato, Luisa R. R. "Quilombos em Matto Grosso: Resistência negra em área de fronteira." In *Liberdade por um fio,* ed. Reis and Gomes. 213–39.

Waldrep, Christopher. "Substituting Law for the Lash: Emancipation and Legal Formalism in a Mississippi County Court." *Journal of American History* 82:4 (1996): 1425–51.

Walker, Daniel E. "No More, No More": Slavery and Cultural Resistance in Havana and New Orleans. Minneapolis: University of Minnesota Press, 2004.

Walvin, James. *Black Ivory: A History of British Slavery.* London: Fontana Press, 1992.

Watkin, E. I. *Roman Catholicism in England from the Reformation to 1950.* London: Oxford University Press, 1957.

Watson, Alan. *Slave Law in the Americas.* Athens: University of Georgia Press, 1990.

Webb, James, Jr. "The Horse and Slave Trade between the Western Sahara and Senegambia." *Journal of African History* 34:2 (1993): 221–46.

White, Gavin. "Firearms in Africa." *Journal of African History* 12:2 (1971): 173–84.

White, Luise. *Speaking with Vampires: Rumor and History in Colonial Africa.* Berkeley: University of California Press, 2000.

Whitman, T. Stephen. *The Price of Freedom: Slavery and Manumission in Baltimore and Early National Maryland.* Lexington: University Press of Kentucky, 1997.

Wickham, Chris. "Gossip and Resistance among the Medieval Peasantry." *Past and Present* 160 (1998): 3–24.

Wurdemann, John G. Notas sobre Cuba. Havana: Letras Cubanas, 1989.

Yacou, Alain. "La présence française dans la partie occidentale de l'île de Cuba au lendemain de la Révolution de Saint-Domingue." Revue française d'histoire d'Outre-Mer 84 (1987): 149–88.

———. "La conspiración de Aponte." Historia y Sociedad 1 (1988): 39–58.

Zeuske, Michael. "The Cimarrón in the Archives: A Re-Reading of Miguel Barnet's Biography of Esteban Montejo." New West Indian Guide/Nieuwe West-Indische Gids 71: 3–4 (1997): 265–79.

———. "Novedades de Esteban Montejo." Revista de Indias 59: 216 (1999): 521–25.

———. "Más novedades de Esteban Montejo." Del Caribe 38 (2002): 95–101.

Index